BASIC soil mechanics

Butterworths BASIC Series includes the following titles:

BASIC aerodynamics
BASIC artificial intelligence
BASIC economics
BASIC hydraulics
BASIC hydrology
BASIC interactive graphics
BASIC investment appraisal
BASIC materials studies
BASIC matrix methods
BASIC mechanical vibrations
BASIC molecular spectroscopy
BASIC numerical mathematics
BASIC operational amplifiers
BASIC soil mechanics
BASIC statistics
BASIC stress analysis
BASIC theory of structures
BASIC thermodynamics and heat transfer

BASIC soil mechanics

G W E Milligan MA, MEng, PhD, CEng, MICE
Department of Engineering Science, University of Oxford
and Fellow of Magdalen College, Oxford

and

G T Houlsby MA, PhD, CEng, MICE
Department of Engineering Science, University of Oxford
and Fellow of Keble College, Oxford

Butterworths
London . Boston . Durban . Singapore . Sydney . Toronto . Wellington

First published 1984
Reprinted 1986

British Library Cataloguing in Publication Data

Milligan, G.W.E.
BASIC soil mechanics.
1. Basic (Computer program language)
2. Soil science—Data processing
I. Title II. Houlsby, G.T.
001.64'24 S598

ISBN 0-408-01365-6

Library of Congress Cataloging in Publication Data

Milligan, G.W.E.
BASIC soil mechanics.

Includes bibliographical references and index.
1. Soil mechanics—Data processing. 2. Basic (Computer program language) I. Houlsby, G. T.
II. Title. III. Title: B.A.S.I.C. soil mechanics.
TA710.M518 1984 624.1'5136'0285424 83-26172
ISBN 0-408-01365-6

Typeset by Illustrated Arts Ltd., Sutton, Surrey
Printed and bound by Mackays of Chatham Ltd.

Preface

The importance of computers and computer programming has grown rapidly in recent years. This growth has been widespread and has been accelerated by the advent of microcomputers. Almost all engineers have, or will soon have, access to computer facilities and it is necessary that they see computing as a natural everyday tool.

Computing is in the curriculum of all university engineering courses and indeed it is often taught in schools. However, many students find it a difficult and alien subject, often because it is taught by computer scientists or mathematicians and not related to their engineering subjects. There are a number of advantages that can accrue from linking the study of engineering and computing. Engineering can help computing by showing its relevance and by providing exercises with which programming techniques can be learned, practised and developed. Computing is then seen as both an essential and natural activity for engineers. Computing can help engineering because a clear exposition (and consequent understanding) of the engineering equations and procedures is required in order to write a successful computer program. This approach can lead to an enthusiastic acceptance of computing. It is of value not only to student engineers but also to practising engineers who may not have been taught computing or who were perhaps initially discouraged by their experiences.

This book uses BASIC, a programming language which is often criticised by computer scientists because of its unstructured nature. For engineers BASIC has many advantages in addition to its widespread availability as a built-in microcomputer language. Its use of simple English statements makes it easy to learn and remember so that an engineer can surmount the initial hurdle of computing and then employ it usefully at infrequent intervals afterwards. BASIC programs can be rapidly developed because special compiling, linking and editing routines are not required — a help to both beginner and expert. Moreover, it allows a student to complete a programming exercise within a short period of time (e.g. an afternoon or study period) and thereby retain his interest and confidence in computing.

BASIC Soil Mechanics aims to help students to become proficient at BASIC programming by using it in an important engineering subject. The large number of short programs contained in this book should convince the reader of the ease and value of computing. Many of the programs will be found to be of immediate use to practising engineers, but the main aim of the book is to demonstrate programming techniques and encourage engineers to develop programs to suit their own problems.

Chapter 1 is an introduction to BASIC. Chapter 2 introduces soil mechanics and the nature of soil as an engineering material. Chapters 3 to 7 cover the topics normally included in elementary texts on soil mechanics: identification, classification and testing of soils; earth pressures; stability of slopes; and foundations. The flow of water through soils is an important aspect of both theoretical and practical geotechnical engineering, and most books on soil mechanics include a chapter on this subject. However, it has already been well covered by the chapter on Seepage in an earlier volume of this series, *BASIC Hydraulics* by P. D. Smith, and to avoid duplication has been omitted from this book.

Each chapter contains a summary of relevant theory, worked examples of computer programs, and a set of problems. The theory content may be useful as an introduction to, or revision synopsis of, the subject but is by no means comprehensive. The examples are posed as questions, solutions to which are given as the listing of a possible program with an example of its output and some 'program notes'. These notes explain the structure of the program and how it uses the theory. The problems involve modifying or extending the example programs and writing completely new ones. The reader can learn about both BASIC programming and soil mechanics by studying the examples and attempting the problems.

The usefulness of many of the programs may be much enhanced by using the computer to plot results in a graphical form, either for display on a screen or as a 'hard copy' on paper. Unfortunately the methods of programming computers for graphical displays tend to be specific to individual systems and examples have not therefore been incorporated in this book. Readers are encouraged, however, to make use of facilities for graphics on the computers they are using as the rewards are well worth the effort involved.

Acknowledgement for assistance is due to the series editors, M. J. Iremonger and P. D. Smith, particularly the latter for providing the initial encouragement to write this book; to Sue Clarke and Sally Schofield for typing the manuscript; and to Judith Takacs for preparing the figures.

G. W. E. Milligan
G. T. Houlsby

Contents

Principal notation

e		Void ratio
n		Porosity
v		Specific volume
w		Water content (moisture content)
S_r		Degree of saturation
G_s		Specific gravity of soil grains
A		Air content
γ		Unit weight (weight of unit volume)
ρ		Density (mass of unit volume)
σ, σ'		Total and effective normal stress } See Note 1
τ		Shear stress } See Note 1
u		Pore pressure
p	=	$\frac{1}{3}(\sigma_1 + \sigma_2 + \sigma_3)$ Mean normal stress
q	=	$(\sigma_1 - \sigma_3)$
s	=	$\frac{1}{2}(\sigma_1 + \sigma_3)$ } For plane-strain conditions
t	=	$\frac{1}{2}(\sigma_1 - \sigma_3)$ } For plane-strain conditions
ε		Direct strain
γ		Shear strain
ε_v	=	$(\varepsilon_1 + \varepsilon_2 + \varepsilon_3)$ Volumetric strain
ε_s	=	$\frac{2}{3}(\varepsilon_1 - \varepsilon_3)$ (triaxial test), $(\varepsilon_1 - \varepsilon_3)$ (plane-strain)
M		Critical state friction constant
ϕ		Angle of internal friction
c		Cohesion
c_u, c_a		Undrained shear strength, adhesion
δ		Angle of wall friction
E		Young's modulus (modulus of elasticity)
G		Shear modulus
ν		Poisson's ratio
A, B		Pore pressure parameters
C_c, C_s		Compression and swelling indices
K_0, K_a, K_p		Earth pressure coefficients
C_v		Coefficient of consolidation
U_t		Degree of consolidation

T_v	Time factor for consolidation
m_v	Coefficient of volume change
F	Factor of safety
N_s	Stability number
m, n	Stability coefficients
r_u	Pore pressure ratio
N_c, N_q, N_γ	Bearing capacity factors
i_c, i_q, i_γ	Inclination factors for bearing capacity

Notes: 1. The normal suffixes used in stress analysis are employed in this book:

x, y, z co-ordinate directions
1, 2, 3 principal values

2. Other suffixes used, unless otherwise noted, have the following meanings:

d	dry soil
sat.	saturated soil
u	undrained
h	horizontal
v	vertical
a	active
p	passive
o	initial or surface value
f	value at failure

3. Notation for forces and geometrical properties is defined in the text and figures.

Chapter 1

Introduction to BASIC

1.1 The BASIC programming language

The programs in this book are written in the BASIC programming language. BASIC, an acronym for Beginner's All-purpose Symbolic Instruction Code, was developed at Dartmouth College, USA, as a simple general purpose language. It was originally intended for use on time-sharing computer systems, but has gained widespread popularity as the main language associated with microcomputers. BASIC is both easy to learn and to use. A program can quickly be written, typed in at the computer, run and corrected and run again if any errors are present. The main disadvantages of simple BASIC relate to its lack of structure (see section 1.4) but this is not an important consideration for short programs such as those in the following chapters.

This book aims to help in the learning of BASIC by applying it to some relevant engineering problems. This aim can be met by the reader studying the examples, possibly copying them and then trying some of the problems. Although this book does not specifically teach the grammar of BASIC, a short description of the simple BASIC used is given in the next section.

1.2 The elements of BASIC

1.2.1 Mathematical expressions

One of the main objects of the example programs in this book is to evaluate the equations that arise in soil mechanics. These equations contain numerical constants (e.g. ϕ', the angle of friction), variables (e.g. x) and functions (e.g. sine). All numbers are treated identically whether they are integer (e.g. 36) or real (e.g. 36.1). An exponential form is used to represent large or small numbers (e.g. 3.61E6 which equals 3.61×10^6). Numerical variables are represented by a letter or a letter followed by another letter or a digit (e.g. E, EA or E1). An operation, such as square root, can be done using a built-in function (e.g. SQR(X)). The argument in brackets (X) can be a number, a variable or a mathematical expression. For trigonometric functions (SIN(X), COS(X), etc) the argument is always interpreted as being measured in radians. Other functions include a natural logarithm and the exponential (LOG and EXP respectively), ABS which selects the absolute value of argument and INT which selects the integer part of the argument.

Mathematical equations also contain operators such as plus and minus, etc. These operators have a hierarchy in that some are performed by the computer before others. In descending order of hierarchy the operators are

to the power of (∧)
multiply (*) and divide (/)
add (+) and subtract (−)

Thus, for example, multiplication is done before addition. The computer works from left to right if the operators have the same hierarchy. Brackets can be used to override any of these operations. Hence $\frac{a+b}{3c}$ may be written as (A + B)/(3*C) or (A + B)/3/C. Although this last expression will be calculated correctly, it is not recommended as it is possibly ambiguous.

In some computers the names of variables are restricted to single letters, but usually variable names of two or more letters are permitted. In order to allow use of variable names which have mnemonic value, two letter names are used in many of the programs in this book.

1.2.2 Assignment statements

A BASIC program is a sequence of statements which define a procedure for the computer to follow. As it follows this procedure the computer allocates values to each of the variables. The values of some of these variables may be specified by data that is input to the program. Others are generated in the program using, for instance, the assignment statement. This has the form

line number [LET] variable = mathematical expression

where the LET is usually optional and therefore omitted. (All optional parts of expressions will be shown enclosed in square brackets thus: [optional]). For example the root of a quadratic equation

$$x_1 = \frac{-b + \sqrt{(b^2 - 4ac)}}{2a}$$

may be obtained from a statement such as

```
100 X1 = (-B + SQR(B∧2 - 4*A*C))/(2*A)
```

It is important to realise that an assignment statement is not itself an equation. It is an instruction to give the variable on the left-hand side the numeric value of the expression on the right-hand side. Thus it is possible to have a statement

```
50 X = X + 1
```

which increases by 1 the value of X.

Each variable can have only one value at any one time unless it is subscripted (see section 1.2.7).

Note that all BASIC statements (i.e. all the program lines) are numbered. This defines the order in which they are executed.

1.2.3 Input statements

For 'interactive' programs the user specifies variables by input of their values when the program is run. The input statement has the form

line number INPUT variable 1 [, variable 2, . . .]

e.g.

```
20 INPUT A,B,C
```

When the program is run the computer prints ? as it reaches this statement and waits for the user to type values for the variables, e.g.

? <u>5,10,15</u>

which makes A = 5, B = 10 and C = 15 in the above example.

Here, and throughout the rest of this book, the input which must be typed in by the programmer (rather than printed by the computer) is indicated by underlining. The programmer of course does not have to underline the input, and the underlining would not appear when normally using a computer.

An alternative form of data input is useful if there are many data or if the data are not to be changed by the user (e.g. see Example 7.5). For this type of data specification there is a statement of the form

line number READ variable 1 [, variable 2, . . .]

e.g.

```
20 READ A,B,C
```

with an associated statement (or number of statements) of the form

line number DATA number 1 [, number 2, . . .]

e.g.

```
1 DATA 5,10,15
```

or

```
1 DATA 5
2 DATA 10
3 DATA 15
```

DATA statements can be placed anywhere in a program — it is often convenient to place them at the beginning or end in order that they can be easily changed (e.g. see Example 4.2).

When using built-in data it is sometimes necessary to read the data from their start more than once during a single program run. This is done using the statement

```
line number RESTORE
```

1.2.4 Output statements

Output of data and the results of calculations, etc. is done using a statement of the form

```
line number PRINT list
```

The list may contain variables or expressions, e.g.

```
200 PRINT A,B,C,A*B/C
```

text enclosed in quotes, e.g.

```
10 PRINT "INPUT A,B,C, IN MM"
```

or mixed text and variables, e.g.

```
300 PRINT "STRESS IS",S,"KN/M∧2"
```

The items in the list are separated by commas or semicolons. Commas give tabulation in columns, each about 15 spaces wide. A semicolon suppresses this spacing and if it is placed at the end of a list it suppresses the line feed. If the list is left unfilled a blank line is printed.

Note the necessity to use PRINT statements in association with both 'run-time' input (to indicate what input is required) and READ/DATA statements (because otherwise the program user has no record of the data).

1.2.5 Conditional statements

It is often necessary to enable a program to take some action if, and only if, some condition is fulfilled. This is done with a statement of the form line number IF expression 1 { conditional operator } expression 2 THEN GOTO line number

where the possible conditional operators are

=	equals
<>	not equal to
<	less than
<=	less than or equal to
>	greater than
>=	greater than or equal to

For example a program could contain the following statements if it is to stop when a zero value of A is input,

```
20 INPUT A
30 IF A <> 0 THEN 50
40 STOP
50 . . .
```

Note the statement

line number STOP

which stops the running of a program.

1.2.6 Loops

There are several means by which a program can repeat some of its procedure; the self-repeating sequence of program statement is called a loop. The simplest such statement is

line number GOTO line number

This can be used, for instance, with the above conditional statement example so that the program continues to request values of A until the user inputs zero.

The most common means of performing loops is with a starting statement of the form

line number FOR variable = expression 1 TO expression 2 [STEP expression 3]

where the STEP is assumed to be unity if omitted. The finish of the loop is signified by a statement

line number NEXT variable

where the same variable is used in both FOR and NEXT statements. Its value should not be changed in the intervening lines.

A loop is used if, for example, *N* sets of data are input by a READ statement and their reciprocals printed, e.g.

```
10 READ N
20 PRINT "NUMBER","RECIPROCAL"
30 FOR I = 1 TO N
40 READ A
50 PRINT A, 1/A
60 NEXT I
```

Loops can also be used to generate data. Consider, for example, a simple temperature conversion program

```
10 PRINT "CENTIGRADE","FAHRENHEIT"
20 FOR C = 0 TO 100 STEP 5
30 PRINT C, 9 * C/5 + 32
40 NEXT C
```

1.2.7 Subscripted variables

It is sometimes very beneficial to allow a single variable to have a number of different values during a single program run (e.g. see Examples 4.3 and 4.4). For instance, if a program contains data for several materials it is convenient for their densities to be called R(1), R(2), R(3), etc. instead of R1, R2, R3, etc. It is then possible for a single statement to perform calculations for all the materials, e.g.

```
50 FOR I = 1 TO N
60 M(I) = V * R(I)
70 NEXT I
```

which determines the mass M(I) for each material from the volume (V) of the body.

A non-subscripted variable has a single value associated with it and if a subscripted variable is used it is necessary to provide space for all the values. This is done with a dimensioning statement of the form

```
line number DIM variable 1 (integer 1) [, variable 2 (integer 2), . . .]
```

e.g.

```
20 DIM R(50), M(50)
```

which allows up to 50 values of R and M. The DIM statement must occur before the subscripted variables are first used.

On some computers it is possible to use a dimension statement of a different form, e.g.

```
20 DIM R(N), M(N)
```

where the value of N has been previously defined. This form, when available, has the advantage of not wasting space. It will not be used in this book.

1.2.8 Subroutines

Sometimes a sequence of statements needs to be accessed more than once in the same program (e.g. see Example 6.3). Instead of merely repeating these statements it is better to put them in a subroutine. The program then contains statements of the form

line number GOSUB line number

When the program reaches this statement it branches (i.e. transfers control) to the second line number. The sequence of statements starting with this second line number ends with a statement

line number RETURN

and the program returns control to the statement immediately after the GOSUB call.

Subroutines can be placed anywhere in the program but it is usually convenient to position them at the end, separate from the main program statements.

Another reason for using a subroutine occurs when a procedure is written which is required in more than one program. It is often desirable to use less common variable names (e.g. X9 instead of X) in such subroutines. This minimizes the possibility of the same variable name being used with a different meaning in separate parts of a program.

Subroutines may also be used to break a program up into logical units, as in Example 7.4.

1.2.9 Other statements

(1) Explanatory remarks or headings which are not to be output can be inserted into a program using

line number REM comment

Any statement beginning with the word REM is ignored by the computer. On some computers it is possible to include remarks on the same line as other statements but this will not be used in this book.
(2) Non-numerical data (e.g. words) can be handled by string variables. A string is a series of characters within quotes, e.g. "STRESS" and a string variable is a letter followed by a $ sign, e.g. S$. They are particularly valuable when printed headings need to be varied.
(3) Multiple branching can be done with statements of the form

```
line
number ON expression THEN line number 1 [, line number 2, . . .]
```

(GOTO is an alternative to THEN in this statement e.g. see Example 7.1) and

```
line
number ON expression GOSUB line number 1 [, line number 2, . . .]
```

When a program reaches one of these statements it branches to line number 1 if the integer value of the expression is 1, to line number 2 if the expression is 2, and so on. An error message is printed if the expression gives a value less than 1 or greater than the number of referenced line numbers.
(4) Functions other than those built into the language such as SIN(X) can be created as defined functions using a DEF statement. For example

```
10 DEF FNA(X) = X∧3 + X∧2 + X + 1
```

defines a cubic function which can be recalled later in the program as FNA (variable) where the value of this variable is substituted for X. A defined function is useful if an algebraic expression is to be evaluated several times in a program (e.g. see Example 7.5).

1.3 Checking programs

Most computers give a clear indication if there are grammatical (syntax) errors in a BASIC program. Program statements can be modified by retyping them correctly or by using special editing procedures. The majority of syntax errors are easy to locate but if a variable has been used with two (or more) different meanings in separate parts of the program some mystifying errors can result.

It is not sufficient for the program to be just grammatically correct. It must also give the correct answers. A program should therefore be checked either by using data which give a known solution or by hand calculation. If the program is to be used with a wide range of data, or by users other than the program writer, it is necessary to check that all parts of it function correctly. It is also important to ensure that the program does not give incorrect but plausible answers when 'nonsense' data is input. It is quite difficult to make programs completely immune to misuse and they become lengthy in so doing. The programs in this book have been kept as short as possible for the purpose of clarity and may not therefore be guaranteed to give sensible results for all input data.

1.4 Different computers and variants of BASIC

The examples in this book use a simple version of BASIC that should work on most computers, even those with small storage capacity. Only single-line statements have been used though many computers allow a number of statements on each line with a separator such as \ or :.

There is one important feature which distinguishes computers, particularly microcomputers, with a visual display unit (VDU). This concerns the number of available columns across each line and the number of rows that are visible on the screen. Simple modifications of some of the programs may be necessary to fit the output to a particular microcomputer.

Various additions to simple BASIC have been made since its inception and these are implemented on a number of computer systems. The programs in this book could be rewritten to take account of some of these advanced features. For example, the ability to use long variable names (e.g. STRESS instead of, say, S or S1) makes it easier to write unambiguous programs. Other advanced facilities include more powerful looping and conditional statements and independent subroutines which make the writing of structured programs easier. In simplistic terms, structured programming involves the compartmentalization of programs and minimises branching due to statements containing 'GOTO line number' and 'THEN line number'. Good program structure is advantageous for long programs.

1.5 Summary of BASIC statements

Assignment

LET	Computes and assigns value
DIM	Allocates space for subscripted variables

Input

INPUT	'Run-time' input of data
READ	Reads data from DATA statements
DATA	Storage area for data
RESTORE	Restores DATA to its start

Output

PRINT	Prints output

Program control

STOP	Stops program run
GOTO	Unconditional branching
IF . . . THEN	Conditional branching
FOR . . . TO . . . STEP	Opens loop
NEXT	Closes loop
GOSUB	Transfers control to subroutine
RETURN	Returns control from subroutine
ON . . . THEN	Multiple branching
ON . . . GOSUB	Multiple subroutine transfer

Comment

REM	Comment in program

Functions

SQR	Square root
SIN	Sine (angle in radians)
COS	Cosine (angle in radians)
ATN	Arctangent (gives angle in radians)
LOG	Natural logarithm (base e)
EXP	Exponential
ABS	Absolute value
INT	Integral value
DEF FN	Defined function

1.6 References

1. Alcock, D., *Illustrating BASIC*, Cambridge University Press, (1977).
2. Kemeny, J. G. and Kurtz, T. E., *BASIC Programming*, John Wiley, (1968).
3. Monro, D. M., *Interactive Computing with BASIC*, Edward Arnold, (1974).
4. Sharp, W. F. and Jacob, N. L., *BASIC, An Introduction to Computer Programming using the BASIC Language*, Free Press, New York, (1979).

Chapter 2

Introduction to soil mechanics

2.1 Aims of soil mechanics

Soil is one of the oldest but still one of the most important materials for civil engineering. Nearly all civil engineering construction is dependent on support provided by the ground, is affected by loading from soil, or is itself built out of soil. As with other engineering materials, it is necessary for the engineer to understand the mechanical behaviour of soil so as to be able to predict under what conditions a construction may fail completely or its deformations become excessive. In addition to strength and stiffness, a third property of soil of great practical importance is its permeability, a measure of the ease with which water may flow through the soil.

Soil is unusual as an engineering material for two reasons. Firstly, it is a natural material which has gone through a complex series of processes during formation and is seldom even approximately homogeneous or isotropic. At any location there will generally be several layers or strata of quite different types of material; within each layer there will be small scale variations in properties. Properties may also vary with orientation; for instance, some soils have thin alternating layers of clay and sand, giving a higher permeability and lower resistance to sliding in a horizontal direction than in a vertical direction. Secondly, the values of engineering properties of soils cover a very wide range. Strength and stiffness can vary over a range of at least three orders of magnitude, and permeability over about ten orders of magnitude. Usually the engineer has to make use of the soil as it exists at the construction site; even when it is used as a construction material he is usually constrained by transport costs to the soil materials available locally. Identification, classification and testing of soils are therefore very important.

The methods of soil mechanics are similar to those of other branches of engineering. Material properties are determined from relatively simple tests, and are used with a physical model of the mechanical behaviour of the material. This model is a sufficiently accurate representation of the complex real behaviour of the material while being simple enough to allow the solution of real engineering problems. The compromise between simplicity and accuracy will change with the importance of the job, while for the reasons given above the application of any theory to real soils is a matter for care and judgement.

2.2 Origin and nature of engineering soil

Working downwards from the earth's surface, an engineer recognizes three main types of material: topsoil, soil and rock. Topsoil includes the weathered humus-containing soil able to support plant life. It is valuable for agriculture and has poor engineering properties and is therefore generally removed at the start of engineering work. Soil and rock are both conglomerates of mineral grains, and the boundary between soil and rock is somewhat arbitrary and often not clear-cut. Soil is basically uncemented and relatively weak, rock cemented and relatively strong. Soil is derived from rocks by processes of chemical and physical weathering, while sedimentary rocks such as shale and sandstone are formed by the compression and cementation of soil materials by geological processes. At least a passing knowledge of geology and geomorphology is essential to a proper understanding of the present distribution and nature of the various types of engineering soil.

Physical processes of weathering, principally the action of ice, water, wind and variations in temperature, retain the original mineral structure of the parent rock but break it down into successively finer pieces. The particles produced are approximately equidimensional and seldom smaller than about 0.002 mm in diameter. The most important mineral is silica which has strong chemical bonds and a stable crystalline packing which is resistant to further degradation. Silica particles form sands and silts.

Chemical processes of weathering, mainly involving water, oxygen and carbon dioxide, alter rock minerals to clay minerals. The latter are very complex, but the most important are built up from basic units of silica and alumina to form rod or sheet structures, which in turn combine in a number of different ways to form more-or-less stable particles. These particles are generally smaller than 0.002 mm, down to colloidal size.

Natural soils are generally a mixture of particles of different sizes. The words clay, silt, sand, gravel, cobbles and boulders are used specifically to define particular ranges of particle sizes, as indicated in Figure 2.1. They are also used more generally to describe natural soils in which the properties of a particular size range predominate. However a 'clay' soil may contain as little as 10 per cent of clay-size particles and seldom more than 50 per cent, but is so described because the presence of clay-size particles has a profound effect on the behaviour of soil. The effect occurs in two main ways which are related to the size of 'pores' in the material and to the different arrangements in which the particles can be packed together.

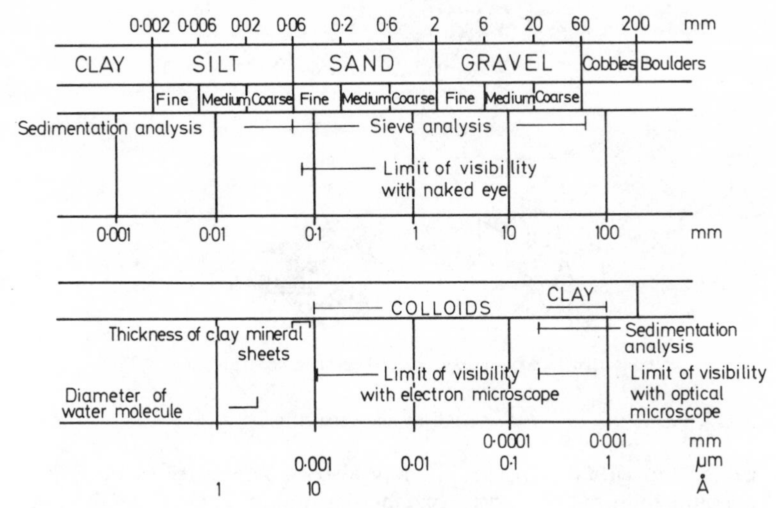

Figure 2.1 Particle sizes

Any particulate material must have spaces or pores between the solid grains, and in soils these are usually filled by water, air or a combination of both. The presence of this pore fluid, and the ease with which it can move through the soil, are of fundamental importance to the behaviour of soils. The size of the pores is obviously related to the size of the particles, and in clay soils will be very small; in mixed soil this will still be true provided there is sufficient clay to fill the spaces between the larger particles. Water can only move through such pores with great difficulty, and the soil will be nearly impermeable; also in very small pores capillary forces may be large, and the resulting 'suction' will hold the material together even when no external stresses are applied. Clay soil therefore appears to be cohesive and plastic, it will hold itself together in a lump and can undergo substantial distortion without breaking up.

The behaviour of soil is greatly affected by the way in which the particles are packed together. The range of states between the loosest and densest possible arrangements of approximately spherical particles is not very great, but with thin clay particles the same grains can occupy very different volumes, and have very different properties, depending on the packing. As a result of differences in origin or stress-history clay particles may exist in any of the arrangements shown in Figure 2.2; in addition some clay particles themselves may change in volume by absorbing or losing molecules of water.

In engineering calculations soil may usually be treated as a continuum, but the material properties are closely related to its particulate nature and its microstructure, the properties and arrangement

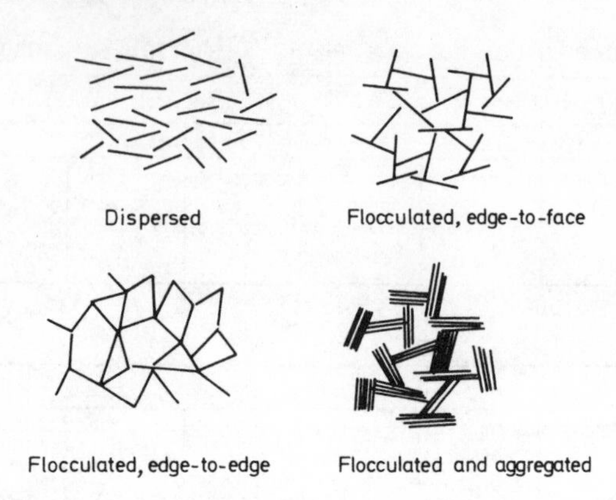

Figure 2.2 Clay microstructures

of the individual particles. It may also be necessary to take into account features of 'fabric' or macrostructure, such as small scale variations in soil type within a major stratum, cracks or fissures, or inclusions of organic materials. Organic soils contain significant quantities of plant or animal remains; peat is a particular example which consists almost entirely of decayed vegetable matter.

2.3 Pore pressures and effective stress

Appreciation of the importance of pore pressures is the first major step towards the understanding of the behaviour of soils. The pore spaces in a soil are interconnected and pressure may therefore be transmitted through the pore fluid between two points in the soil. At any point in an unsaturated soil both pore-air and pore-water pressures may exist, and may be different due to surface tension effects. In a saturated soil there will only be pore-water pressure. If the water is stationary the pressures will vary hydrostatically, increasing with depth below the ground water table. Above the water table the soil may still be saturated by capillary action but pore pressures will be negative (all pressures relative to atmospheric pressure). For water to flow through the soil additional pressure differentials must exist.

The main effect of pore pressure is to reduce the contact forces between soil particles by 'pushing apart' the particles. A better way of expressing this is to say that the total stresses within a body of soil, which must of course satisfy equilibrium conditions within and at the boundaries of the soil mass, are carried partly by fluid pressure in the pores and partly by the contact forces between particles. The deformation and failure of soil are related only to the interparticle forces, which are most conveniently described in terms of an 'effective' stress which is the summation over unit area of the components of interparticle forces normal to any plane under consideration. Algebraically the definition (originally by K. Terzaghi) of effective stress is given by

$$\sigma' = \sigma - u \qquad (2.1)$$

where σ is the total normal stress on some plane at a point within the soil, u is the pore pressure at the point, and σ' is the corresponding normal effective stress on the plane. Fluid cannot resist shear stresses so that the shear stress on a plane is not affected by pore pressure and total and effective values are equal.

Note that effective normal stresses can never be measured directly, as they are the cumulative effect of a large number of local highly stressed contacts between grains, but are determined only as the difference between the total stress and the pore pressure. Equation (2.1) is easily applied in saturated soils in which the pore pressure at a point has a unique value. Its application to unsaturated soils is not considered in this book.

2.4 Notation and units

A list of the notation and units used for the main parameters in soil mechanics appears at the beginning of the book. Parameters are also defined where they first appear in the text. SI units are used throughout. In this system the basic unit of force is the newton (N), of length the metre (m), and of time the second (s). So that reasonable numbers are obtained for realistic values, forces are normally expressed in terms of the larger units kN (10^3N) or MN (10^6N). Similarly stresses are normally expressed in kN/m^2 and unit weights in kN/m^3. Elsewhere the unit N/m^2 is sometimes called the pascal (Pa), so that kPa may be written for kN/m^2. Time-dependent processes in soils often occur very slowly and it is more convenient to use a time unit of a year (yr) than of a second; one year is approximately 31.6×10^6 seconds.

2.5 The scope of this book

Chapter 1 introduced BASIC as a programming language, and this chapter has introduced soil as an engineering material. Chapter 3 deals with methods of identifying and classifying soils, and introduces some of the terminology of soil mechanics. Chapter 4 covers the laboratory and field tests used to determine the material properties for use in engineering calculations. Chapters 5, 6 and 7 deal with three important types of problem encountered in soil mechanics: the determination of earth pressures against retaining structures, the analysis of the stability of slopes, and the design of foundations. Certain other types of problem, such as soil pressures on tunnels and the soil mechanics aspect of highway design, are considered to be beyond the scope of this book.

Each of the following chapters contains a summary of the relevant theory with some discussion of the assumptions, reasoning and limitations involved. The coverage of each topic is necessarily brief; this book does not provide a full course in soil mechanics nor discuss any of the practical problems involved in geotechnical engineering. More detailed coverage may be obtained from the references listed in the following section.

It is also assumed that the reader has some familiarity with the basic concepts of solid mechanics although the particular notation used in soil mechanics is introduced in Chapter 4.

2.6 References

1. Bell, F. G., *Engineering Properties of Soils and Rocks*, 2nd Ed., Butterworths, (1983).
2. BS 1377, *Methods of Test for Soils for Civil Engineering Purposes*, British Standards Institution, (1975).
3. BS 5930, *Code of Practice for Site Investigations*, British Standards Institution, (1981).
4. Craig, R. F., *Soil Mechanics*, 2nd Ed., Van Nostrand Reinhold, (1978).
5. Scott, C. R., *An Introduction to Soil Mechanics and Foundations*, Applied Science Publishers, (1969).
6. Smith, G. N., *Elements of Soil Mechanics for Civil and Mining Engineers*, 5th Ed., Granada, (1982).
7. Atkinson, J. H. and Bransby, P. L., *The Mechanics of Soils*, McGraw-Hill, (1978).
8. Atkinson, J. H., *Foundations and Slopes*, McGraw-Hill, (1981).
9. Bolton, M., *A Guide to Soil Mechanics*, Macmillan, (1979).
10. Poulos, H. G. and Davis, E. H., *Elastic Solutions for Soil and Rock Mechanics*, John Wiley, (1974).

References 1, 2 and 3 provide additional background reading for Chapters 2, 3 and 4. References 4, 5 and 6 are popular standard textbooks on soil mechanics; each covers most of the theory contained in this book in considerably greater detail. In recent years 'critical state' soil mechanics has provided a unified approach to the understanding of soil behaviour. The concepts are introduced in this book but are discussed in much greater detail in Reference 7, while applications to practical problems in soil mechanics are covered by References 8 and 9. Reference 10 provides a very useful compilation of solutions based on elastic theory which may have applications in soil and rock mechanics, and is particularly appropriate to Chapter 7.

Chapter 3

Phase relationships and index tests

ESSENTIAL THEORY

3.1 Phase relationships

As discussed in the previous chapter, soil has three constituent phases: solid (mineral), liquid (usually water) and gas (usually air). It is now necessary to define the relationships between the phases more precisely and introduce some of the standard terminology of soil mechanics.

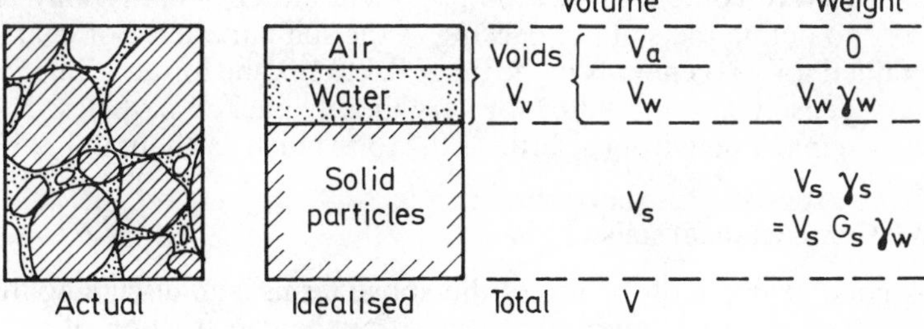

Figure 3.1 Soil phases

A given volume of soil may be represented by the idealization shown in Figure 3.1, and the following terms defined:

Porosity n = volume of voids/total volume (3.1)

Void ratio e = volume of voids/volume of solids $= n/(1-n)$ (3.2)

Specific volume v = total volume containing unit volume of solids

$= 1+e$ (3.3)

Degree of saturation S_r = volume of water/volume of voids (3.4)
= 0 for dry soil
= 1 for saturated soil

Air content A = volume of air/total volume (3.5)

The above relationships are based on volumes. By introducing the unit weight of water γ_w and the specific gravity of the soil grains G_s, relationships based on weights may also be determined:

$$\text{Unit weight } \gamma = \text{total weight/total volume} = [\gamma_w(G_s + S_r.e)]/(1 + e) \quad (3.6)$$

$$\text{from which } \gamma_{sat} = [\gamma_w(G_s+e)]/(1+e) \text{ for saturated soil} \quad (3.7)$$

$$\text{and } \gamma_d = (\gamma_w.G_s)/(1+e) \text{ for dry soil} \quad (3.8)$$

$$\text{Moisture (or water) content } w = \text{weight of water/weight of solids}$$

$$= S_r.e/G_s \quad (3.9)$$

$$= 0 \text{ for dry soil}$$

$$= e/G_s \text{ for saturated soil} \quad (3.10)$$

Each of these interrelated quantities may be used in a particular application. Note that soil is commonly fully saturated in its natural state and the void ratio for a particular soil is then directly related to its moisture content (Equation 3.10) and either quantity may be used to define the state of packing of the soil particles. For coarse grained soil e is generally in the range 0.3 to 1.0, the former value for very densely and the latter for very loosely packed particles. For clays e may range from as little as 0.2 to as much as about 9.

3.2 Compaction of soils

A good example of the use of the above terms is in discussing the compaction of unsaturated soil used for the construction of road bases or embankments. Compaction means the packing together more closely of the grains by mechanical action—rolling, hammering or vibration — by the expulsion of air from the voids, the moisture content remaining approximately constant.

The relative ease with which different soils may be compacted is determined by field trials or laboratory tests in which specimens of soil with several different moisture contents are compacted in a standard manner (see Reference 2). The degree of compaction achieved is measured by the dry density ρ_d of the soil, analogous to the dry unit weight defined above but expressed in units of mass rather than weight. This is a measure of the amount of solid material in unit volume and may be determined from the measured bulk density ρ and the moisture content w using the relationship

$$\rho = (1+w)\rho_d \quad (3.11)$$

It is found that the greatest dry density is obtained at an optimum moisture content, typical results being shown in Figure 3.2. In the same figure curves are plotted showing values of the percentage air content determined from the relationship

$$\rho_d = [G_s(1-A)\rho_w]/(1+w.G_s) \quad (3.12)$$

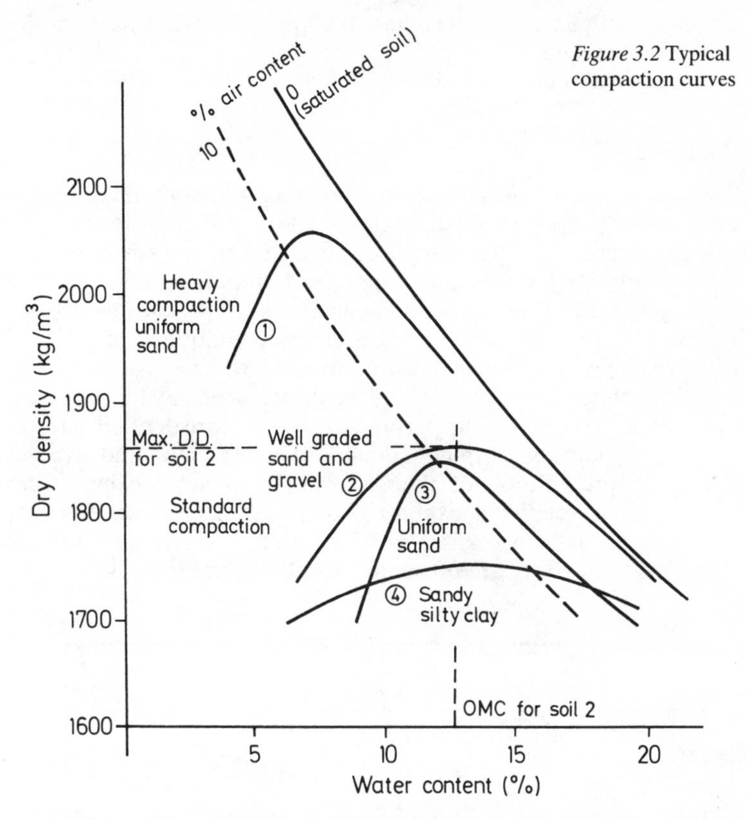

Figure 3.2 Typical compaction curves

in which ρ_w is the density of water. The zero air content (complete saturation) curve obviously forms an upper boundary to the compaction curve on the 'wet' side of optimum.

3.3 Identification and classification of soils

The first stage of an engineering calculation concerning soil is a proper identification of the materials involved. A site investigation usually starts with a survey of information available from geological maps and memoirs, adjacent construction and other local knowledge. Boreholes or trial pits will then be excavated to determine the succession of soil strata and obtain samples of soil for identification and for testing in the laboratory. Tests may also be performed to determine the properties of the soil in situ.

Preliminary classification and description of soils is done in the field on the basis of visual and tactile properties. Standard methods of description, and some simple tests to help distinguish between soils of different types, are given in Reference 3. The purpose of such classification is to provide a quick means of identification of the different soil strata and give a general indication to an experienced engineer of their expected engineering performance. Classification is greatly assisted by the performance of some simple laboratory tests known as index tests, details of which are given in Reference 2.

3.4 Index tests

The first of these is the particle size analysis, to determine the range and relative proportions of particles of different sizes within the soil. For coarse grained soils a sample of soil is shaken or washed through successively finer sieves and the proportion passing through each determined. For fine grained soils sieving is not practicable and the test uses instead the different rates of sedimentation of particles of different sizes in accordance with Stokes' Law. This strictly applies to spherical particles, whereas soil particles, especially those of clay, are not truly spherical. The particle size is therefore defined in terms of an equivalent sphere. The sedimentation test is slow and involves a lengthy preparation of the soil sample to achieve consistent results, so is relatively expensive to perform and consequently sparingly used. Some examples of the way in which the results of particle size analyses are presented are shown in Figure 3.3.

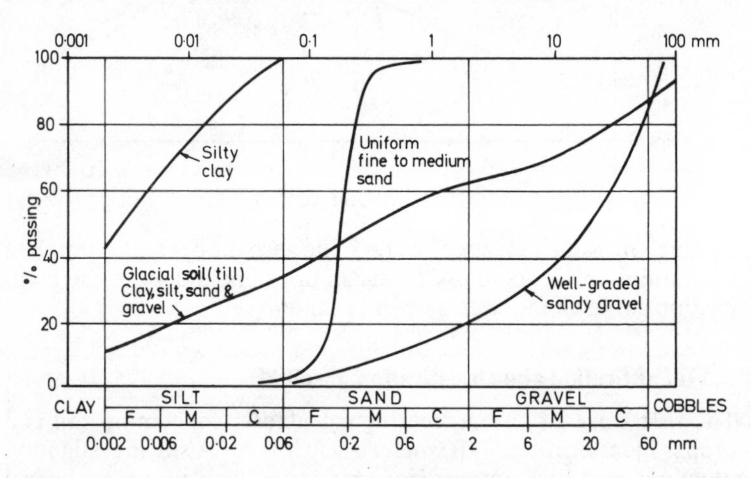

Figure 3.3 Typical grain size distributions

For fine grained soils the determination of the natural moisture content of the soil is of greater use. The moisture contents corresponding to two arbitrarily defined limits of consistency are also determined. The liquid limit (LL) defines the change in behaviour as the soil is made progressively wetter from that of a plastic solid to that of a viscous liquid, and is defined in terms of a standard penetration test. The plastic limit (PL) corresponds to the change from a plastic to a brittle solid as the soil is made progressively drier, and is defined in terms of the ability of the soil to be rolled out into a thin thread without breaking up. Moisture content, liquid limit and plastic limit are expressed as percentages, the latter two to the nearest whole number. The difference between the liquid and plastic limits is known as the plasticity index (PI). LL, PL and PI all tend to increase with the proportion of clay size particles in a soil, and also depend on the nature of the clay minerals. On the basis of investigations of many different soils Casagrande developed a plasticity chart as an aid to the classification of fine grained soils (see Figure 3.4). Note in particular the separation between clay and silty soils by the diagonal line known as the A line. For further details of a standardized system of classification see Reference 3.

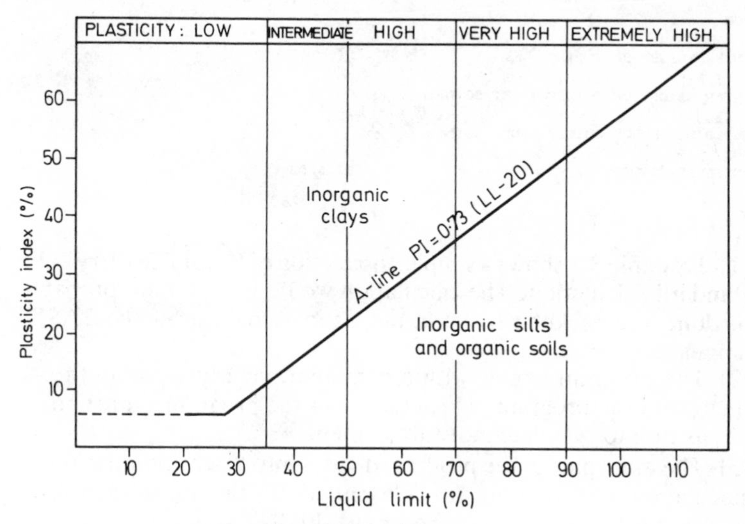

Figure 3.4 Plasticity chart

Natural soil in the ground will usually have a moisture content somewhere from just below the PL to just above the LL; the former value would generally apply to a strong, stiff soil, the latter to a very weak and compressible soil. The relative consistency is given a numerical value known as the liquidity index (LI) and defined by

$$LI = (w - PL)/(PI) \tag{3.13}$$

w being the natural moisture content.

WORKED EXAMPLES

Example 3.1 Determination of moisture content of specimen of soil

Write a program to determine the moisture content of a specimen of soil from the measurements made in the laboratory: the initial weight of a specimen of soil in a container, and the final weight after drying of the specimen in the oven.

```
LIST

10      REM EXAMPLE 3.1. CALCULATE MOISTURE CONTENT
20      REM
30      PRINT "ENTER WEIGHT OF BOTTLE:"
40      INPUT B
50      PRINT "ENTER WEIGHT OF BOTTLE + WET SOIL:"
60      INPUT W
70      PRINT "ENTER WEIGHT OF BOTTLE + DRY SOIL:"
80      INPUT D
90      M = 100 * (W - D)/(D - B)
100     PRINT "MOISTURE CONTENT (PERCENT) = ",M
110     STOP
120     END

READY

RUN

ENTER WEIGHT OF BOTTLE:
? 11.275
ENTER WEIGHT OF BOTTLE + WET SOIL:
? 26.462
ENTER WEIGHT OF BOTTLE + DRY SOIL:
? 22.810
MOISTURE CONTENT (PERCENT) =              31.6602
```

Program notes

(1) Example 3.1 shows a simple application of BASIC to carry out a standard calculation. The calculation would in fact more probably be done on a pocket calculator, but it serves to illustrate the BASIC language.
(2) The program begins with a comment, clearly separated from the rest of the program, which describes the program's function — in this case to calculate moisture content.
(3) The main part of the program divides into three sections: input, calculation and output. In lines 30 to 80 the three quantities required for the calculation are input. PRINT statements are used to prompt for the input in an unambiguous manner. Line 90 is the only line of calculation required following the definition of moisture content in Equation (3.9) and in line 100 the result is output, again using the PRINT statement to make the output clear. Lines 110 and 120 are required to stop execution and mark the end of the BASIC program.
(4) It is tempting to try to shorten programs by omitting some of the apparently unnecessary PRINT and REM statements, but it is bad programming practice to do so. Comments which make a program clearer to the reader and user are always worthwhile.

Example 3.2 Calculation of basic soil properties

Write a program to calculate the unit weight, moisture content, voids ratio, and degree of saturation of a sample of soil of known volume. The weight of the sample before and after drying have been measured, and the specific gravity of the soil particles either assumed or measured in a separate test.

```
LIST

10      REM EXAMPLE 3.2. CALCULATE BULK AND DRY DENSITIES, VOIDS RATIO ETC.
20      REM
30      PRINT "ENTER VOLUME OF SOIL (M3):"
40      INPUT V
50      PRINT "ENTER WEIGHT OF WET SOIL (KGF):"
60      INPUT WW
70      PRINT "ENTER WEIGHT OF DRY SOIL (KGF):"
80      INPUT WD
90      PRINT "ENTER SPECIFIC GRAVITY OF PARTICLES:"
100     INPUT G
110     B = 0.00981 * WW / V
120     PRINT "BULK UNIT WEIGHT (KN/M3)          = ",B
130     D = 0.00981 * WD / V
140     PRINT "DRY UNIT WEIGHT (KN/M3)           = ",D
150     M = 100 * (WW - WD) / WD
160     PRINT "MOISTURE CONTENT (PERCENT)        = ",M
170     E = (G * 9.81 / D) - 1
180     PRINT "VOIDS RATIO                       = ",E
190     S = M * G / E
200     PRINT "DEGREE OF SATURATION (PERCENT) = ",S
210     PRINT "ENTER 0 TO END, 1 FOR NEW DATA:"
220     INPUT I
230     IF I=1 GOTO 30
240     STOP
250     END

READY

RUN

ENTER VOLUME OF SOIL (M3):
? .01437
ENTER WEIGHT OF WET SOIL (KGF):
? 28.756
ENTER WEIGHT OF DRY SOIL (KGF):
? 23.820
ENTER SPECIFIC GRAVITY OF PARTICLES:
? 2.65
BULK UNIT WEIGHT (KN/M3)          =         19.6309
DRY UNIT WEIGHT (KN/M3)           =         16.2613
MOISTURE CONTENT (PERCENT)        =         20.7221
VOIDS RATIO                       =         .598678
DEGREE OF SATURATION (PERCENT) =            91.7247
ENTER 0 TO END, 1 FOR NEW DATA:
? 1
ENTER VOLUME OF SOIL (M3):
? .01466
ENTER WEIGHT OF WET SOIL (KGF):
? 29.032
ENTER WEIGHT OF DRY SOIL (KGF):
? 24.010
ENTER SPECIFIC GRAVITY OF PARTICLES:
? 2.65
BULK UNIT WEIGHT (KN/M3)          =         19.4273
DRY UNIT WEIGHT (KN/M3)           =         16.0667
MOISTURE CONTENT (PERCENT)        =         20.9163
VOIDS RATIO                       =         .618034
DEGREE OF SATURATION (PERCENT) =            89.6846
ENTER 0 TO END, 1 FOR NEW DATA:
? 0
```

Program notes

(1) The program in Example 3.2 is another example of a simple repetitive calculation. This time several quantities (bulk and dry unit weight, moisture content, voids ratio and degree of saturation) are calculated and output.

(2) For convenience of reading the program the calculation and output parts of the program are interleaved in lines 110 to 200; this is a method which is often useful — each result is output immediately after it is calculated.
(3) Lines 210 to 230 introduce a simple control structure. If many calculations are to be carried out it is convenient to offer the user a choice after each calculation of either stopping or continuing with another calculation. The program prompts for a value of the control variable I, and if it is 1 returns to the beginning of the program for a new calculation. This feature will be included in many of the other programs in this book.

```
LIST

10      REM EXAMPLE 3.3. CALCULATE VERTICAL TOTAL AND EFFECTIVE STRESSES
20      REM
30      DIM G(20),Z(20)
40      PRINT "ENTER GROUND WATER LEVEL (M BELOW SURFACE):"
50      INPUT W
60      K=0
70      PRINT "ENTER DEPTH AND BULK UNIT WEIGHT (M, KN/M3) (0, 0 TO END):"
80      K=K+1
90      INPUT Z(K),G(K)
100     IF Z(K)<>0 GOTO 80
110     K=K-1
120     PRINT " "
130     PRINT "Z","SIG-V(TOT)","SIG-V(EFF)","U"
140     PRINT "(M)","(KN/M2)","(KN/M2)","(KN/M2)"
150     PRINT " "
160     PRINT 0,0,0,0
170     T = 0
180     ZL = 0
190     FOR I=1 TO K
200     T = T + (Z(I) - ZL) * G(I)
210     ZL = Z(I)
220     U = 9.81 * (Z(I) - W)
230     IF U<0 THEN U = 0
240     E = T - U
250     PRINT Z(I),T,E,U
260     NEXT I
270     STOP
280     END

READY

RUN

ENTER GROUND WATER LEVEL (M BELOW SURFACE):
? 3.2
ENTER DEPTH AND BULK UNIT WEIGHT (M, KN/M3) (0, 0 TO END):
? 2, 14.75
? 4.3, 17.6
? 5.9, 17.2
? 9.7, 19.83
? 15.6, 18.9
? 0, 0

Z               SIG-V(TOT)      SIG-V(EFF)      U
(M)             (KN/M2)         (KN/M2)         (KN/M2)

 0              0               0               0
 2              29.5            29.5            0
 4.3            69.98           59.189          10.791
 5.9            97.5            71.013          26.487
 9.7            172.854         109.089         63.765
 15.6           284.364         162.72          121.644
```

Example 3.3 Calculation of vertical total and effective stresses

Write a program to interpret the information on the ground water level and bulk densities of successive strata of soil as determined from a site investigation borehole, so as to calculate the vertical total and effective stresses within the ground.

Program notes

(1) Example 3.3 introduces several new features. Its purpose is to calculate the vertical total and effective stresses at different levels in a soil, given a set of bulk densities for layers between particular depths, and also the level of the water table. Since several depths (the levels of the boundaries between layers) are required they will be stored in an array Z(I), and the corresponding unit weights in an array G(I). Z and G are declared in a DIM statement at the beginning of the program. The unit weight G(I) will apply to the soil between depth Z(I – 1) and Z(I).
(2) Since the number of layers is not specified in advance the procedure in lines 60 to 110 is used to input the values of Z and G. K is a counting variable which is first initialized to zero and then increased by 1 in line 80. The first time line 90 is executed it therefore reads Z(1), G(1). Line 100 tests the value of Z and if it is not zero returns to 80, so that K is now set to 2 and Z(2), G(2) are read. This continues until a zero value is given for Z(K). At this stage K will be set to one more than the number of layers, line 110 resets it to the number of layers.
(3) Lines 120 to 160 write the headings for a table and the first line — zero stresses at zero depth. Using the variables T and ZL for the total weight of the soil in the layers above and the level of the last layer, the FOR . . . NEXT loop in 190 to 260 calculates the stresses at the level of each layer. Line 190 sets up the loop to count through each layer, and 200 adds the weight of the current layer (Z(I) – ZL)*G(I) to T to give the new value of T — the total stress at the bottom of the layer. ZL is then reset to its new value. The pore pressure U is calculated from the depth below the water table in 220, and 230 checks for the case where the point is above the water table in which case U is zero. Line 240 calculates the vertical effective stress using the Terzaghi relationship and 250 outputs the results.

Example 3.4 Analysis of particle size distribution

Write a program to analyse the results of a sieve analysis of a coarse grained soil. The measurements made during the test are the dry weights of the material retained on each of a successively finer set of sieves as the sample of soil is shaken through the stack of sieves. The results should be suitable for plotting in the form shown in Figure 3.3.

```
LIST

10      REM EXAMPLE 3.4. ANALYSIS OF PARTICLE SIZE DISTRIBUTION
20      REM
30      DIM S(10),T(10),W(10)
40      N=10
50      S(1)=37.5
60      S(2)=20
70      S(3)=14
80      S(4)=6.3
90      S(5)=3.35
100     S(6)=1.18
110     S(7)=0.6
120     S(8)=0.3
130     S(9)=0.212
140     S(10)=0.063
150     PRINT "PARTICLE SIZE DISTRIBUTION"
160     PRINT "ENTER WEIGHTS RETAINED ON SIEVES:"
170     FOR I=1 TO N
180     PRINT "SIEVE SIZE (MM): ",S(I)
190     INPUT W(I)
200     NEXT I
210     PRINT "WEIGHT IN COLLECTOR:"
220     INPUT T(N)
230     REM
240     REM CALCULATE WEIGHT PASSING SIEVES
250     REM
260     FOR I=N-1 TO 1 STEP -1
270     T(I)=T(I+1) + W(I+1)
280     NEXT I
290     WT=T(1) + W(1)
300     REM
310     REM CALCULATE PERCENTAGES
320     REM
330     PRINT "SIEVE (MM)     PERCENT PASSING"
340     FOR I=1 TO N
350     T(I)=100 * T(I) / WT
360     PRINT S(I),T(I)
370     NEXT I
380     PRINT "ENTER 0 TO STOP, 1 FOR MORE DATA"
390     INPUT J
400     IF J=1 GOTO 160
410     STOP
420     END

READY

RUN

PARTICLE SIZE DISTRIBUTION
ENTER WEIGHTS RETAINED ON SIEVES:
SIEVE SIZE (MM):              37.5
? 60.2
SIEVE SIZE (MM):              20
? 36.8
SIEVE SIZE (MM):              14
? 41
SIEVE SIZE (MM):              6.3
? 38.2
SIEVE SIZE (MM):              3.35
? 25.7
SIEVE SIZE (MM):              1.18
? 18.6
SIEVE SIZE (MM):              .6
? 17.6
SIEVE SIZE (MM):              .3
? 25.4
SIEVE SIZE (MM):              .212
```

```
? 27
SIEVE SIZE (MM):              .063
? 19.3
WEIGHT IN COLLECTOR:
? 39.4
SIEVE (MM)      PERCENT PASSING
 37.5           82.7606
 20             72.2222
 14             60.4811
 6.3            49.5418
 3.35           42.1821
 1.18           36.8557
 .6             31.8156
 .3             24.5418
 .212           16.8099
 .063           11.2829
ENTER 0 TO STOP, 1 FOR MORE DATA
? 0
```

Program notes

(1) The program in Example 3.4 calculates the percentages passing a set of different sieve sizes, given the weights of material retained on each of the sieves. The example is given for 10 sieves, but this could easily be changed. The arrays S(I), W(I) and T(I) are used for the sieve size, weight retained on the sieve, and total weight passing a given sieve.
(2) Line 40 sets N to the number of sieves — all the rest of the program uses N rather than 10 so that if the number of sieves is changed then only lines 30 and 40 need be altered. Lines 50 to 140 simply set up the sizes of sieves being used. Lines 170 to 200 make use of a FOR . . . NEXT loop to input the weights on the sieves.
(3) In 260 to 280 the weight passing each sieve is calculated, in this case STEP −1 is used in the FOR . . . NEXT loop since the weight passing is calculated by going backwards through the list of sieves, totting up the total weight T(I) passing sieve I by adding the weight W(I + 1) on sieve I + 1 to the weight T(I + 1) passing sieve I + 1. The weight passing the smallest sieve must of course first be input at lines 210 to 220.
(4) Line 290 calculates the total weight and 330 to 370 output the results — again using a FOR . . . NEXT loop to calculate percentages. Finally the routine for selecting another set of data is included.

Example 3.5 Statistical analysis of test results

In civil engineering construction a very large number of tests are often carried out on materials to check that they comply with specified standards. Because soil is a rather variable material tests on a single type of soil may show considerable scatter, so that rather than rejecting material on the grounds of isolated test results it is better to carry out a statistical analysis of tests. Write a program to calculate the mean and standard deviation from a series of results. The program should allow the possibility of adding new data to a long list of results without the whole list needing to be retyped.

```
LIST

10      REM EXAMPLE 3.5. CALCULATION OF MEAN AND STANDARD DEVIATION OF
20      REM A SET OF MEASURED QUANTITIES
30      REM
40      PRINT "ENTER 1 FOR NEW SET OF DATA"
50      PRINT "       2 FOR ADDITION TO PREVIOUS SET OF DATA"
60      INPUT I
70      ON I GOTO 80,120
80      NL=0
90      TL=0
100     SL=0
110     GOTO 140
120     PRINT "ENTER THREE QUANTITIES RECORDED FROM PREVIOUS DATA SET:"
130     INPUT NL,TL,SL
140     REM
150     REM ENTER THE NEW SET OF DATA
160     REM
170     PRINT "ENTER EACH NEW READING, -999 TO END:"
180     NN=0
190     TN=0
200     SN=0
210     NN=NN+1
220     INPUT X
230     IF X=-999 GOTO 270
240     TN=TN+X
250     SN=SN+X*X
260     GOTO 210
270     NN=NN-1
280     IF I=1 GOTO 340
290     N=NL
300     T=TL
310     S=SL
320     PRINT "RESULTS OF PREVIOUS SET OF READINGS:"
330     GOSUB 1000
340     N=NN
350     T=TN
360     S=SN
370     PRINT "RESULTS OF THIS SET OF READINGS:"
380     GOSUB 1000
390     IF I=1 GOTO 450
400     N=NL+NN
410     T=TL+TN
420     S=SL+SN
430     PRINT "RESULTS OF TOTAL READINGS:"
440     GOSUB 1000
450     PRINT "ENTER 1 TO STOP"
460     PRINT "       2 TO ADD MORE READINGS TO THIS SET"
470     PRINT "       3 FOR NEW SET OF DATA"
480     INPUT J
490     ON J GOTO 500,530,580
500     PRINT "IF MORE ANALYSES ARE TO BE CARRIED OUT RECORD THESE NUMBERS:"
510     PRINT N,T,S
520     STOP
530     NL=N
540     TL=T
550     SL=S
560     I=2
```

```
570     GOTO 140
580     PRINT "IF MORE ANALYSES ARE TO BE CARRIED OUT RECORD THESE NUMBERS:"
590     PRINT N,T,S
600     GOTO 40
1000    M = T/N
1010    SD = SQR(S/N - M*M)
1020    PRINT " "
1030    PRINT "READINGS:";N,"MEAN:";M,"STANDARD DEV.:";SD
1040    PRINT "PROBABLE RANGE:",M-2*SD,"TO",M+2*SD
1050    PRINT " "
1060    RETURN
1070    END

READY

RUN

ENTER 1 FOR NEW SET OF DATA
      2 FOR ADDITION TO PREVIOUS SET OF DATA
? 1
ENTER EACH NEW READING, -999 TO END:
? 1.86
? 3.45
? 2.87
? 2.83
? 1.89
? -999
RESULTS OF THIS SET OF READINGS:

READINGS: 5   MEAN: 2.58    STANDARD DEV.: .616117
PROBABLE RANGE:             1.34777      TO            3.81223

ENTER 1 TO STOP
      2 TO ADD MORE READINGS TO THIS SET
      3 FOR NEW SET OF DATA
? 2
ENTER EACH NEW READING, -999 TO END:
? 3.56
? 2.87
? 3.41
? 1.98
? 3.12
? -999
RESULTS OF PREVIOUS SET OF READINGS:

READINGS: 5   MEAN: 2.58    STANDARD DEV.: .616117
PROBABLE RANGE:             1.34777      TO            3.81223

RESULTS OF THIS SET OF READINGS:

READINGS: 5   MEAN: 2.988   STANDARD DEV.: .557257
PROBABLE RANGE:             1.87349      TO            4.10251

RESULTS OF TOTAL READINGS:

READINGS: 10  MEAN: 2.784   STANDARD DEV.: .621838
PROBABLE RANGE:             1.54032      TO            4.02768

ENTER 1 TO STOP
      2 TO ADD MORE READINGS TO THIS SET
      3 FOR NEW SET OF DATA
? 1
IF MORE ANALYSES ARE TO BE CARRIED OUT RECORD THESE NUMBERS:
 10           27.84         81.3734
```

Program notes

(1) The program starts by offering the user the choice of entering a new set of data or adding more information to a previous set of data. In order to calculate the mean and standard deviation of a set of readings x_i only three quantities are required: the number of readings n, the sum x_i and the sum of the squares x_i^2. If the set of data is a continuation from a previous set then these three quantities for the previous data must be entered.
(2) The program again uses a special number (−999) to indicate the end of the input data.
(3) The program outputs the mean and standard deviation for each of the previous set of data (if it exists), the current set of data and the total set. This can be convenient if the user wishes to see how some new readings compare with previous values. This repetition of the calculation is best achieved by using the subroutine at lines 1000 to 1070. Note the use of a separate set of line numbers to make the subroutine clearly separated from the rest of the program.
(4) The mean and standard deviation are calculated in the usual way:

$$m = \frac{\Sigma x_i}{n} \quad \text{and} \quad s = \sqrt{\frac{\Sigma x_i^2}{n} - m^2}$$

where m is the mean and s the standard deviation. The summations are made as the data is input at lines 240 and 250. The output routine prints also $m - 2s$ and $m + 2s$, which gives the probable range for most readings.
(5) Note that the variables N, T and S are used in the subroutine. These are set to NN, TN and SN for the new readings, NL, TL and SL for the previous set and the sum for the combined set. When the calculation is finished the values are output so that they can be recorded for re-input to the program if more data is to be added later.

PROBLEMS

(3.1) Modify the program in Example 3.1 to allow the option of inputting several sets of data in succession.
(3.2) Modify the program in Example 3.2 for an application in fill control: if the moisture content is above 32% or below 28% a suitable warning message should be printed, and similarly if the degree of saturation is below 95%.
(3.3) The program in Example 3.3 gives a set of vertical stresses at different levels which could be used to plot σ'_v against depth. If the ground water table is not at the level of one of the layers the plot will not be quite exact in the layer containing the water table. Modify the

program in Example 3.3 to give the value of the stresses at the water table — inserted at the correct point in the table of stresses.

(3.4) Write a program to calculate vertical effective stress at any arbitrary depth, given information similar to that used in Example 3.3.

(3.5) Rewrite the program in Example 3.4 to input the sieve sizes using DATA and READ statements.

(3.6) Modify the program in Example 3.4 to check whether the percentages passing each of the sieves fall within a specified range — the upper and lower bounds on the percentage for each sieve being stored in arrays U(I) and L(I). If any sieve is out of specification a warning should be printed.

Chapter 4

Tests for strength and compressibility of soils

ESSENTIAL THEORY

4.1 Introduction and notation

As mentioned in Chapter 2, the geotechnical engineer is concerned with three main properties of soil: its strength, compressibility and permeability. Problems involving permeability are not included in this book as they are covered in the companion volume *BASIC Hydraulics* by P. D. Smith. In this chapter the mechanical properties of soil, and the tests used to determine them, are reviewed. When soil is loaded, either by its own weight or by external forces, a measure of its strength is needed to ensure that there is an adequate factor of safety against complete failure. A measure of compressibility is also needed to ensure that the deformations under working conditions are not excessive.

This chapter involves discussion of the stresses and strains occurring in the soil, and it is necessary to clarify the notation used. The usual symbols σ and τ are used for normal and shear stresses respectively acting on a particular plane within the soil. Principal stresses, acting on planes on which shear stresses are zero, are denoted by σ_1, σ_2 and σ_3. Three dimensional states of stress are usually described in terms of stress invariants p and q where

$$p = \tfrac{1}{3}(\sigma_1 + \sigma_2 + \sigma_3) \tag{4.1}$$

$$q = (\sigma_1 - \sigma_3) \tag{4.2}$$

p being a measure of the mean normal stress on a soil element, q of the maximum shear stress on any plane in the element.

In two dimensional plane strain conditions (in which $\epsilon_2 = 0$) the influence of the usually unknown intermediate principal stress σ_2 is not included and the parameters used are s and t where

$$s = \tfrac{1}{2}(\sigma_1 + \sigma_3) \tag{4.3}$$

$$t = \tfrac{1}{2}(\sigma_1 - \sigma_3) \tag{4.4}$$

When using Mohr's circle to define a state of stress, note that s defines the centre of the circle and t its radius. In all cases the use of a prime (′) will be used to indicate effective stresses, as in Equation (2.1).

Strains may be similarly expressed in terms of volumetric and shear components ϵ_v and ϵ_s defined in terms of the principal strains ϵ_1, ϵ_2 and ϵ_3 by

$$\epsilon_v = (\epsilon_1 + \epsilon_2 + \epsilon_3) \tag{4.5}$$

$$\epsilon_s = \tfrac{2}{3}(\epsilon_1 - \epsilon_3) \tag{4.6}$$

Under plane-strain conditions $\epsilon_2 = 0$ and $\epsilon_v = (\epsilon_1 + \epsilon_3)$, and in this case ϵ_s is re-defined as $\epsilon_s = (\epsilon_1 - \epsilon_3)$.

4.2 Failure of soil in shear

The failure of soil is essentially controlled by friction, failure occurring on any plane within the soil whenever the shear stress exceeds a certain proportion of the effective normal stress

$$\tau' = \mu\sigma' = \sigma' \tan \phi' \tag{4.7}$$

It must be emphasised that this relationship holds in terms of the effective stresses, the part of the stresses carried by the soil 'skeleton'—the solid particles in contact with each other. The constant of proportionality μ is commonly expressed in the form $\mu = \tan \phi'$, where ϕ' is known as the angle of internal friction of the soil.

Certain heavily compressed or partly cemented soils exhibit a resistance to shear stress even when the normal stress is zero. The above relationship then takes the form

$$\tau' = c' + \sigma' \tan \phi' \tag{4.8}$$

If the state of stress at a point in a soil mass is represented by Mohr's circle, the condition for failure occurs when a circle first touches the line given by Equation (4.7) or (4.8), as circle 1 in Figure 4.1. Circle 2 would represent a state of stress at which failure was not

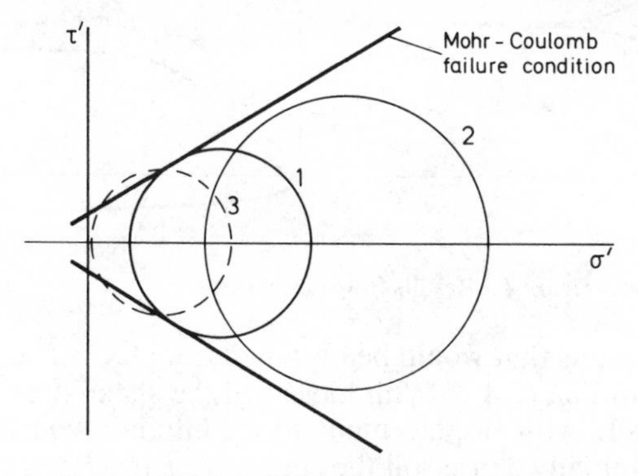

Figure 4.1 Mohr's circles of stress states in soil

was not occurring, while circle 3 would represent a state of stress which could not occur as failure would already have taken place. The failure condition is referred to as the Mohr-Coulomb criterion.

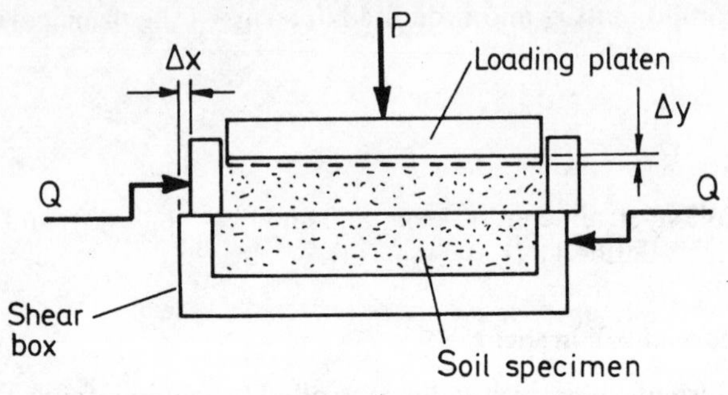

Figure 4.2 Apparatus for direct shear test

The Mohr-Coulomb criterion can be demonstrated by the direct shear test, carried out in a shear box, the arrangement of which is shown diagrammatically in Figure 4.2. The soil is forced to shear on a particular plane under a known normal stress, and the shear stress required determined from the horizontal force applied. The stresses on the failure plane as calculated from the boundary forces are total stresses; as effective stresses are needed the pore pressure must either be measured and Equation (2.1) used, or any pore pressures generated during the test must be allowed to dissipate so that total and effective stresses are equal. It is extremely difficult to measure pore pressures in a direct shear test; for this reason, and because conditions of stress and strain within the soil are not uniform, the test is of limited use. However it is often performed on coarse grained soils because sample preparation is simple, while the high permeability of the soil allows pore pressures to dissipate quickly.

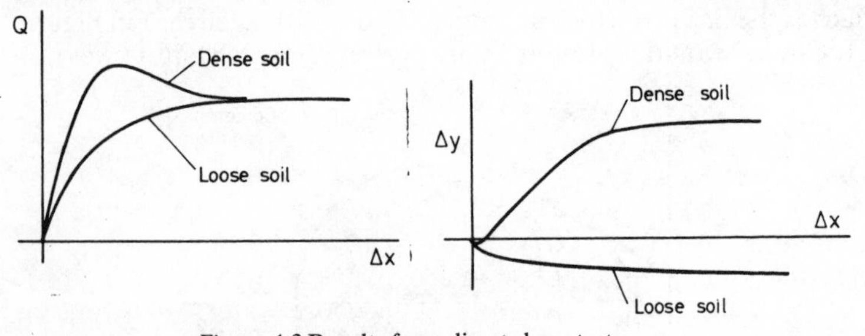

Figure 4.3 Results from direct shear test

Typical results that would be obtained from loose and dense soil are shown in Figure 4.3. With loose soil the shear stress increases monotonically with displacement to an ultimate constant value, while with initially dense soil the shear stress reaches a peak value before dropping to approximately the same ultimate value. The value of ϕ' corresponding to the ultimate condition depends on the mineralogy, roughness and grading of the soil particles. The peak value depends in addition on the initial density and the stress level, being greatest when the soil is initially very dense and the stress level is low. Note that the peak value coincides with the most rapid rate of rise in the top half of the shear box, which indicates that the soil is expanding while being sheared. At the end of both tests the soil is undergoing shear without further change in volume.

The significance of volume change or dilation during shearing of soil is discussed further in section 4.4. One point that may be noted now is the difference in behaviour under increasing loads of large masses of loose and dense soil due to the different stress-strain curves. In dense soil, small local variations in stress or initial density will allow parts of the soil to reach peak strength first. These parts of the soil will then start to get weaker with further shear distortion, while the soil around is still pre-peak and therefore getting stronger. Further deformation will tend to concentrate into these weakening zones, which develop into relatively narrow failure 'planes' separating and allowing relative movement between non-deforming blocks of soil. In initially loose soil this strain-softening behaviour is absent and there is less tendency for deformation to concentrate into such narrow zones.

4.3 Compression of soils and the oedometer

The compression of soils by stresses less than those required to cause failure is important because it may give rise to deformations of the soil which directly affect engineering structures, and also because the strength of soil is affected by its density and will increase with compression.

If an element of saturated soil is acted upon by an isotropic increment of stress in which $\Delta\sigma = \Delta\sigma_2 = \Delta\sigma_3$, the immediate effect will be for the stress to be carried by the pore-water which is much stiffer than the soil skeleton, thus $\Delta u = \Delta p$, $\Delta p' = 0$. If drainage is now permitted water will flow from the soil until $\Delta u = 0$, $\Delta p' = \Delta p$ and the element is again in equilibrium under increased effective stresses and at a reduced specific volume or void ratio. It is found experimentally that e (or v) reduces approximately linearly with the logarithm of p'. The process by which the specific volume of a soil is reduced due to squeezing out of the pore-water by increasing stress is known as consolidation; it is not to be confused with compaction as discussed in section 3.2.

Of more immediate interest to engineers is the case of one dimensional consolidation, which means the compression in one direction only (usually vertical) of a layer of soil by an increase in the stress in that direction while strains in the perpendicular (horizontal) directions are prevented. Such an effect occurs when soil first forms by deposition from a suspension, each element of soil consolidating under the weight of soil deposited above it. It may also occur during construction if a layer of fill is placed over a soil stratum of large lateral extent, and is an approximation to the situation beneath the middle of a large foundation, embankment or other superficial loading.

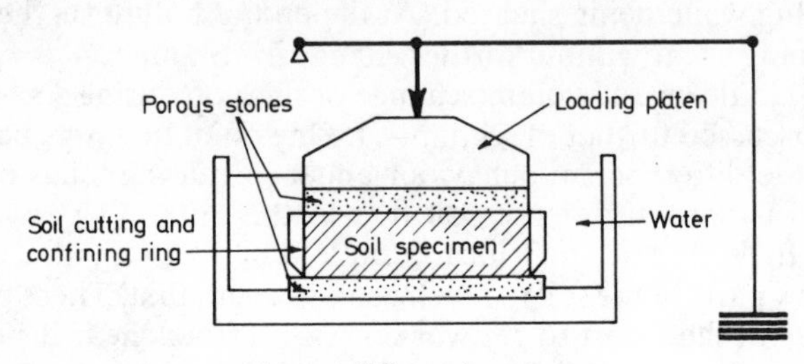

Figure 4.4 The oedometer

One dimensional compression is modelled in the laboratory in a consolidation test using an apparatus called an oedometer, shown diagrammatically in Figure 4.4. The vertical stress is increased in increments, the excess pore pressure in each increment allowed to dissipate over a sufficient time interval, and the void ratio at the end of each stage determined (see worked Example 4.2). The test is normally performed on specimens of saturated fine grained soils.

For soil which is being loaded for the first time to stress levels greater than it has previously experienced the void ratio is found to reduce approximately linearly with the logarithm of σ_v'. Such a soil is said to be normally consolidated, and the line joining successive points representing such states (ABCD in Figure 4.5) is known as the 'virgin compression (or consolidation) line'. If a soil element is unloaded, either by a natural process such as erosion of overlying sediment or as a result of a construction process such as excavation, it increases in volume again but at a very much lower rate as shown by typical 'swelling lines' BE and CF in Figure 4.5. A soil represented by a state such as E or F is said to be overconsolidated. On reloading the soil it returns along the swelling line before continuing down the virgin compression line, along a path such as FCD. Test curves for specimens of real soil usually show some hysteresis in the unloading/reloading cycle and a transition curve rather than a sharp elbow at the junction between reloading and virgin compression lines.

For an overconsolidated soil such as at E the highest stress which it has experienced (at B) is known as its preconsolidation pressure. The ratio of σ'_B to σ'_E is known as the overconsolidation ratio (OCR). For stresses below the preconsolidation pressure the soil will be relatively stiff, volume changes will be recoverable on unloading and the behaviour may be represented over a limited

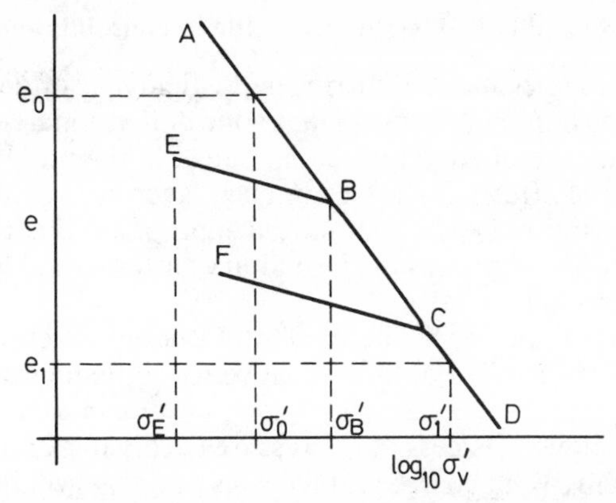

Figure 4.5 e-$\log_{10}\sigma'_v$ relationship for one dimensional compression

range of stress by linear elasticity. Once the preconsolidation pressure is exceeded the rate of deformation increases considerably and volume changes are not fully recoverable.

Consolidation test results are usually plotted in terms of $\log_{10}\sigma'_v$ as in Figure 4.5. The slope of the virgin compression line is then known as the Compression Index (C_c), so that

$$C_c = \frac{e_0 - e_1}{\log_{10}(\sigma_1'/\sigma_0')} \tag{4.9}$$

The slope of a swelling line such as BE is known as the Swelling Index (C_s) and is defined similarly.

Results may alternatively be plotted in terms of $\ln p'$ since

$$p' = \frac{(1 + 2K_0)\,\sigma'_v}{3} \tag{4.10}$$

K_0 is the ratio of horizontal to vertical stresses and is approximately constant for one dimensional loading, being usually estimated as

$$K_0 = (1 - \sin\phi') \tag{4.11}$$

The slopes of the virgin compression and swelling lines in such a plot are then usually given the symbols λ and κ.

Note also that

$$q' = (1 - K_0)\,\sigma'_v > 0 \tag{4.12}$$

Thus one dimensional consolidation involves shear as well as volumetric distortion of the soil.

4.4 Magnitude and rate of settlements due to consolidation

The amount of settlement occurring in the field may often be calculated by approximating the problem to one dimensional conditions. The soil strata are divided into a number of horizontal layers; the initial and final effective stresses in each layer are estimated; the change in thickness of each layer for the appropriate change in stress is determined using the results of oedometer tests; and the results are summed for all the layers.

In addition to calculating the final total amount of settlement, an estimate is often needed of the rate at which settlement will occur. This will depend on the ease with which water can drain from the loaded soil, allowing excess pore pressures to dissipate and consolidation to occur. With coarse grained soils drainage will be so rapid that consolidation settlements will usually occur during the course of construction and in any case will probably be fairly small. With fine grained soils consolidation settlements may continue for months or years and have a considerable effect on the performance of the completed works.

Terzaghi developed a simple theory for one dimensional consolidation which is reasonably accurate for saturated homogeneous soils provided the changes in stress and subsequent strains are relatively small. It relates the excess pore pressure (u), the depth (z) below the top of a layer of soil, and the time (t) after the application of an increment of stress, by the differential equation

$$\frac{\partial u}{\partial t} = C_v \frac{\partial^2 u}{\partial z^2} \quad (4.13)$$

C_v is the coefficient of consolidation. It is a function of the compressibility and permeability of the soil and is assumed to be constant for a particular soil. Common units for C_v are m^2/year. Note that Equation (4.13) is in terms of the excess pore pressure, that is the increase in pore pressure from an initial equilibrium pore pressure, due to the increment in stress. In the long term the pore pressure will return to its equilibrium value and the excess pore pressure will dwindle to zero.

Solutions of Equation (4.13) describing the progress of consolidation are conveniently expressed in terms of a dimensionless time factor T_v and the average degree of consolidation U_t (also dimensionless) at time t, where

$$T_v = C_v t / H^2 \quad (4.14)$$

$$U_t = \rho_t / \rho_\infty \quad (4.15)$$

H is the thickness of the soil layer, ρ_t the settlement at time t and

ρ_∞ the final settlement once consolidation is complete. If H is taken to be the thickness of a soil layer from which pore pressures may dissipate only by drainage to the top surface, the same analysis may be used when drainage can also occur to a permeable underlying stratum by invoking symmetry and considering H as half the thickness of the layer (see Figure 4.6).

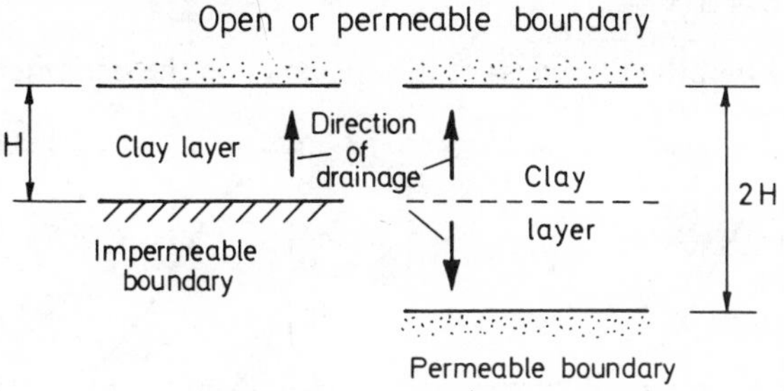

Figure 4.6 Boundary conditions of clay layers

The relationship between U_t and T_v depends somewhat on the initial distribution of excess pore pressures; results for three cases are given in Figure 4.7. In most cases the initial excess pore pressures will be approximately uniform, but under certain conditions one of the triangular distributions may be a better approximation.

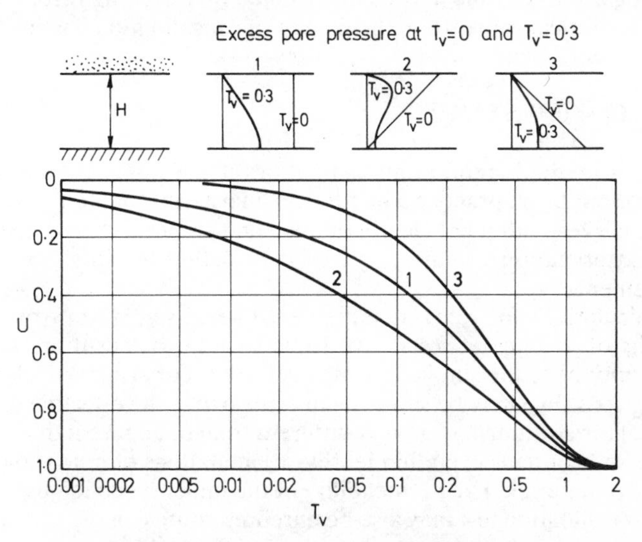

Figure 4.7 U–T_v relationships for different initial conditions

To allow calculations to be made of the progress of consolidation it is necessary to obtain a value of C_v for the soil. This is usually done by back analysis from the settlement curve for each increment of loading in the oedometer using the same one dimensional consolidation theory. Two methods are in common use. The first makes use of the fact that during the initial stages of consolidation the settlement is proportional to the square root of time. Results are plotted as shown in Figure 4.8(a) and C_v determined from

$$C_v = \pi H^2/4t, \tag{4.16}$$

H being half the thickness of the specimen in the oedometer.

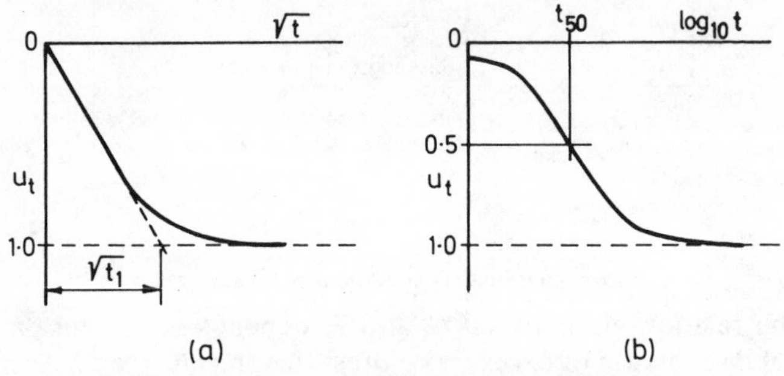

Figure 4.8 Interpretation of consolidation curves by (a) root time method (b) log time method

Alternatively results are plotted as shown in Figure 4.8(b) against the logarithm of time and C_v determined by matching results with the theoretical value when the degree of consolidation has reached 50 per cent, then

$$C_v = 0.196(H^2/t_{50}) \tag{4.17}$$

Both methods require an estimation of the final settlement to determine U_t; in practice a horizontal final asymptote to the settlement curve is often not clearly defined and empirical constructions for extrapolation of curves are used to define the final value of settlement.

Calculations for rates of settlement of real soil strata from the results of oedometer tests have to be treated with caution. Basic assumptions of the simple one dimensional theory (in particular C_v being constant) may be seriously in error, while the effective thickness of the stratum may be very different from its apparent thickness if the soil contains even thin lenses or laminations of much coarser grained material. The permeability of the intact specimen used for the consolidation test may also be unrepresentative of the soil in the ground, where drainage may be aided by structural features such as fissures or root holes.

4.5 Triaxial testing

Triaxial testing has been developed to allow investigation of the stress-strain behaviour of soils under conditions in which the stresses are reasonably uniform and the magnitudes of the principal stresses controlled. The apparatus used is shown diagrammatically

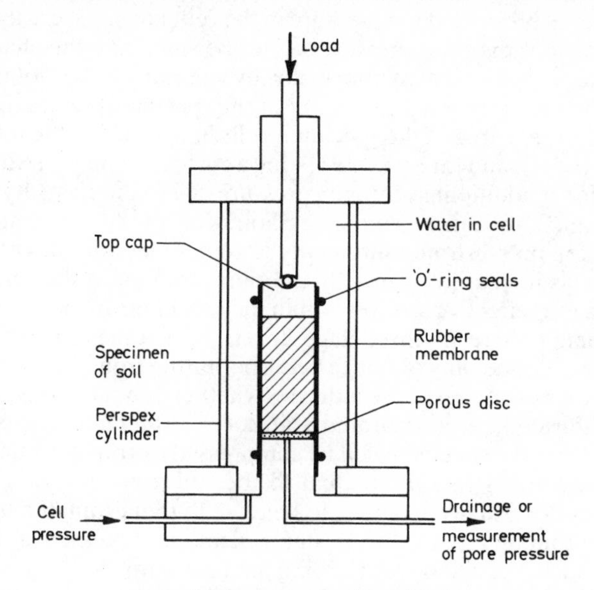

Figure 4.9 Apparatus for triaxial tests

in Figure 4.9. The cylindrical specimen, commonly 75 mm long and 38 mm in diameter but larger for more accurate testing, is enclosed in a rubber membrane and placed in a fluid-filled cell. By pressurizing the fluid in the cell a hydrostatic stress system is applied to the specimen, while a deviator stress may be applied through rigid end-caps and a loading piston. It is assumed that the horizontal principal stresses throughout the sample are equal to the cell pressure, (P_c), and the vertical principal stress (σ_1) is given by

$$\sigma_1 = P/A + p_c \tag{4.18}$$

where P is the load in the piston and A the cross-sectional area of the specimen.

Any type of soil may be tested. Specimens of cohesive fine grained soils are obtained by cutting or turning from larger 'undisturbed' samples of soil; samples of coarse grained soils cannot usually be retrieved intact from the field and specimens are usually made by reconstituting the soil in a former of the required size. Good practice requires the use of lubricated ends to the specimens to reduce shear stresses at the ends and maintain uniform stresses throughout the specimen. Arrangements are made, usually nowadays by electrical transducers, to control and monitor the displacement of the piston and the pressure in the cell, and measure the load in the piston, the pore pressure in the specimen and the change in volume of the specimen (normally by measuring the volume of water sucked into or squeezed out of the specimen). Note that the cross-sectional area of the specimen will change during the test (see Example 4.3), and for accurate testing a correction may also have to be made for additional confining pressure due to the elasticity of the membrane. To ensure complete saturation of the specimen the datum for pressure measurements is sometimes raised; the pore pressure is increased by exactly the same amount as the cell pressure, so that effective stresses within the specimen are not changed.

In general there are three stages to a test. First the cell pressure is increased without any drainage being permitted from the specimen. The pore pressure response indicates whether or not the specimen is fully saturated; in a saturated soil $\Delta u = \Delta p$. Drainage is then allowed and the specimen will be compressed isotropically until the mean effective stress p' is equal to the cell pressure. This initial stress level is usually arranged to be close to that estimated to exist in the ground. Finally the deviator stress is increased until the sample fails, the cell pressure being held constant.

During this third stage, the specimen may either be 'drained' or 'undrained'. In drained tests the load is increased slowly enough for any excess pore pressures developed to dissipate, so that the effective stresses in the specimen are always equal to the applied total stresses. Water may drain into or out of the specimen and its volume change accordingly. In undrained tests no drainage is allowed, and since both the pore water and the soil particles are practically incompressible the specimen may change shape but not volume. Pore pressures will be generated during shearing of the specimen, and effective stresses will no longer equal total stresses. The test may be performed more quickly than a drained test but pore pressures must be measured to allow the effective stresses to be determined.

4.6 Test results and the critical state model

Results are usually plotted in terms of deviator stress and pore pressure or volumetric strain against axial strain. Typical results are shown in Figure 4.10; the difference in behaviour between loose and dense soils mentioned in section 4.2 is again observed. Dense soils, including dense sands and heavily overconsolidated clays, tend to

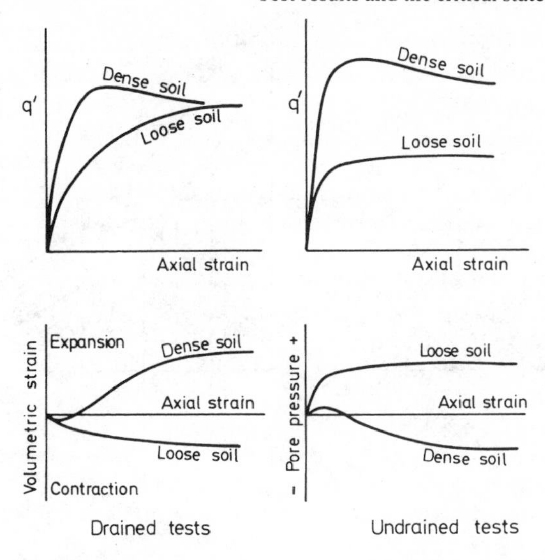

Figure 4.10 Typical results from triaxial tests

expand on shearing in a drained test, or develop negative pore pressures in an undrained test. Loose soils, such as normally consolidated clay or very loose sands, show volume reduction or generate positive pore pressures. It has been found that a soil tends on shearing towards a particular specific volume or void ratio which depends only on the mean effective stress p' and is independent of the initial void ratio. This 'critical' void ratio is found to vary with the logarithm of p'; all possible critical states plot on an $e/\ln p'$ graph as a straight line parallel to the isotropic and virgin one dimensional compression lines (see Figure 4.11(a)). This explains the difference in behaviour of loose and dense soils. A loose soil, starting to the right of the critical state line at point A and tested under undrained conditions so that the void ratio cannot change, must follow path AB to reach a final state at B (Figure 4.11(b)). This involves a reduction in p' even though the total stresses are increasing, which can only occur by an increase in pore pressure. With a dense soil starting at C the opposite occurs; whether pore pressure will be positive or negative will depend on the relative rates of increase in p and p'.

It is also illuminating to plot the stress paths for the tests on a graph of q against p and p'. In a triaxial test $\Delta q = \Delta\sigma_1$ and $\Delta p = \frac{1}{3}\Delta\sigma_1$, since σ_2 and σ_3 are constant. Hence the path of total stress is always a straight line of gradient 3. In a drained test the effective stresses equal the total stresses and follow the same line. In an undrained test the effective stress path differs from the total stress path by a distance parallel to the p-axis equal to the pore pressure.

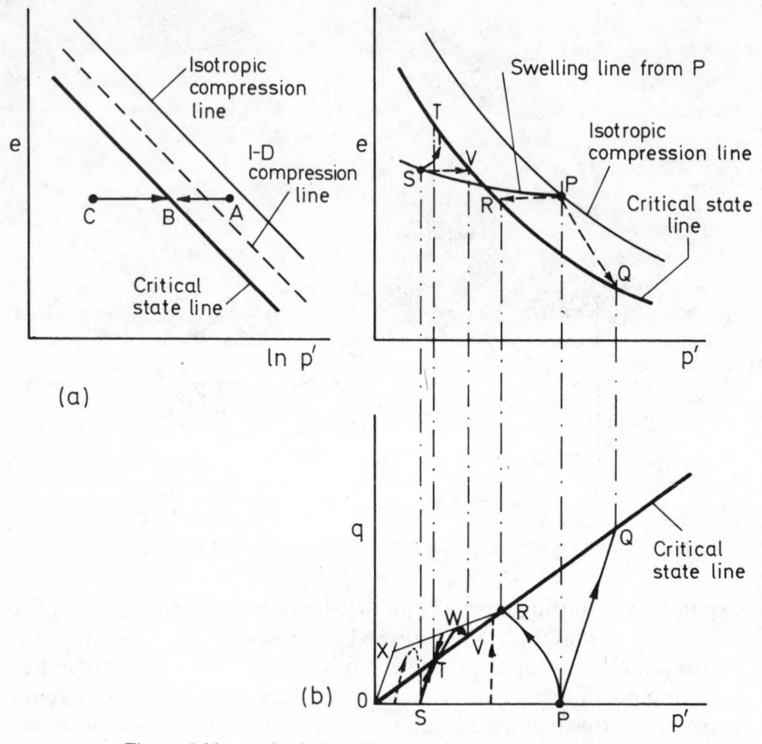

Figure 4.11 e-q-p′ relationships and the critical state line

A typical effective stress path for a normally consolidated clay in an undrained test is shown in Figure 4.11(b) by PR for a specimen initially consolidated to P. PQ would be the total stress path and hence also the effective stress path in a drained test on the same specimen. If the specimen was overconsolidated by compression to P and then unloading to S, the total stress path during shearing would be ST and a typical effective stress path in an undrained test would be SV.

Again the final 'critical' states are found to be approximately on a single line, through the origin, the gradient of which is given the symbol M. In fact the critical state line is a unique line in $e-q-p'$ space, defining states at which deformation takes place at a constant ratio of q/p' and a constant void ratio which is a logarithmic function of p'. The lines in Figure 4.11(b) are the projections of this unique line onto the e/p' and q/p' planes. The points P – V in Figure 4.11(b) may be transferred from the q/p' plot, showing at least qualitatively how the compression and shearing behaviour of soil are linked by this critical state model.

Dense soils are observed to reach a peak value of deviator stress (and hence q) before dropping to the critical state value; the effective stress path therefore rises above the critical state line to a peak value at W before falling to V. Further stress paths for soils of other overconsolidation ratios are shown dotted in Figure 4.11(b). It can be seen that the line RP (for normally and lightly overconsolidated soils) and RX (for heavily overconsolidated soils) provide limits to the effective stress paths of specimens with preconsolidation pressure P tested under undrained conditions.

Further discussion of the critical state interpretation of soil behaviour is outside the scope of this book, but is covered fully in Reference 7. Briefly, it is argued that RP and RX are sections in a q/p' plane of three dimensional surfaces (known as the Roscoe and Hvorslev surfaces respectively) in $e - q - p'$ space defining boundaries to possible soil states; that drained stress paths also lie within or on these state boundary surfaces; that the soil behaves more or less elastically within the state boundaries, but yields or fails when it reaches them. Description of stress-shear strain behaviour can also be included to provide a complete although fairly complex model of soil behaviour. In practice it is still a considerable simplification of the behaviour of real soils, but is very useful for its unification of all aspects of soil behaviour.

The angle of internal friction ϕ_c' in the critical state is related to M by

$$\sin \phi'_c = \frac{3M}{(M+6)} \qquad (4.19)$$

Note that stress paths are sometimes plotted in terms of $\frac{1}{2}(\sigma_1 - \sigma_3)$ against $\frac{1}{2}(\sigma_1 + \sigma_3)$, or s against t in the notation of this book (although sometimes confusingly referred to as q and p). Such notation should really be reserved for plane-strain conditions, when

$$\sin \phi'_c = t'/s' \qquad (4.20)$$

In either case

$$\sin \phi'_c = \frac{(\sigma_1' - \sigma_3')}{(\sigma_1' + \sigma_3')} \qquad (4.21)$$

from which

$$\frac{\sigma_1'}{\sigma_3'} = \frac{1 + \sin \phi'}{1 - \sin \phi'} \qquad (4.22)$$

For design purposes the critical state value of ϕ' is clearly the maximum safe value that can be used for loose soils; for dense soils which have a peak strength greater than at the critical state the use of ϕ'_c may be unduly conservative. Note that the value of ϕ' at peak, found from the slope of the line OW in Figure 4.11(b), will depend on the stress history of the soil element under consideration. It will also generally be higher under plane-strain conditions than measured in the triaxial test, due to the extra restraint against dilation under plane-strain conditions.

4.7 Cam clay

One of the more important aspects of critical state theory is that it can be used to construct complete mathematical models of soil behaviour, describing the response of the soil under any changes of stress and strain. One such model is the 'Cam clay' model of soil behaviour, which is particularly suitable for describing clays under combined conditions of consolidation and shearing.

A full description of the Cam clay model is beyond the scope of this book, but an outline of the main equations follows. For more detail see Reference 7, Chapter 13. The Cam clay model is an elastic-plastic one in which increments of strain are divided into elastic and plastic components, so that

$$d\varepsilon_v = d\varepsilon_v{}^e + d\varepsilon_v{}^p \tag{4.23}$$

$$d\varepsilon_s = d\varepsilon_s{}^e + d\varepsilon_s{}^p \tag{4.24}$$

The elastic strains are small and reversible if the stresses are removed, but the plastic strains are irrecoverable and tend to be larger. As long as the stress point remains within the yield surface plotted in $p' - q$ space then the strains are purely elastic and $d\varepsilon_v{}^p = d\varepsilon_s{}^p = 0$. The shape of the yield surface is assumed to be given by

$$q = Mp' \ln(p_c / p') \tag{4.25}$$

as shown in Figure 4.12. M is the frictional constant introduced in the last section, but p_c may vary and depends on the stresses which have been applied to the soil in the past.

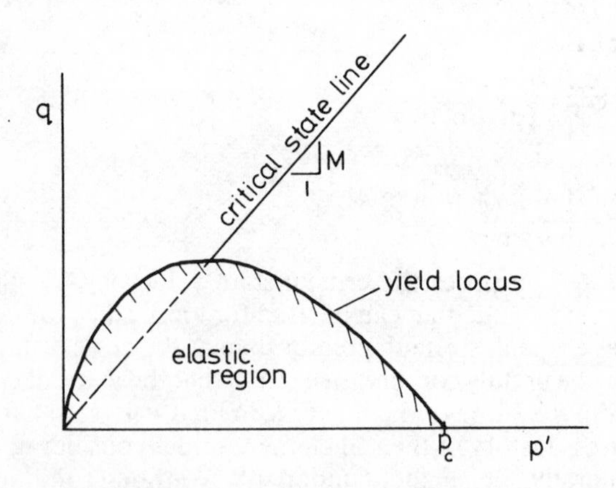

Figure 4.12 The yield locus for Cam clay

The elastic volumetric strain may be determined from the slope of a swelling line in $\ln p'$-v space (see section 4.3) and is given by

$$d\varepsilon_v^{\,e} = \kappa \frac{dp'}{vp'} \tag{4.26}$$

κ being a constant for a given soil.

The elastic shear strain may then be determined, assuming a constant Poisson's ratio ν, from

$$d\varepsilon_s^{\,e} = \frac{2\kappa\,(1+\nu)\,dq}{9\,vp'(1-2\nu)} \tag{4.27}$$

The values of the specific volume v may be determined from the formula

$$v = N - \lambda \ln p_c - \kappa \ln(p'/p_c) \tag{4.28}$$

where N is the value of v for the soil isotropically normally consolidated at $p' = 1.0\ \text{kN/m}^2$ and λ is the slope of the virgin compression line (see section 4.3).

When the combined values of q and p' are such that the stress point reaches the yield surface then both elastic and plastic strains occur. The derivation of the plastic strains is more complex, but it can be shown that

$$d\varepsilon_v^{\,p} = (\lambda-\kappa)\,\frac{[(M-q/p')dp' + dq]}{Mp'v} \tag{4.29}$$

$$d\varepsilon_s^{\,p} = \frac{d\varepsilon_v^{\,p}}{(M-q/p')} \tag{4.30}$$

At this stage the value of p_c increases so that the stress point always remains on the yield surface, which expands, with p_c being determined from Equation (4.25). If at any stage the stress changes are such that the stress point moves back within the yield surface then only elastic strains occur; the yield surface remains at its new enlarged size, defined by the new value of p_c' until the stresses are such that they again reach the yield surface.

All the above applies only for the case where $q < Mp'$. If at any stage q exceeds Mp' the behaviour of the soil is more complex and is not discussed here. The above formulae cannot be used to obtain explicit values for the strains due to arbitrary changes in stress, but may be used *incrementally*. The changes of stress are divided into a number of small increments dp' and dq and the formulae used to derive $d\varepsilon_v$ and $d\varepsilon_s$. This procedure is of course best carried out by a computer.

4.8 Undrained shear strengths

A sample of soil removed from the ground will lose the confining stresses it experienced there. However if the soil is fine grained and impermeable there will be negligible drainage from it, its volume cannot change and as the soil skeleton attempts to expand negative pore pressure (suctions) will be induced which will maintain the existing mean effective stress.

Placing a specimen of such soil in a triaxial cell and applying a cell pressure without allowing drainage will increase the total stress within the specimen but not the effective stress since any increase in confining pressure will be balanced by an equal increase in pore pressure. When a deviator stress is applied and failure induced, still without drainage occurring, the specimen will fail at the same value of maximum shear stress whatever the cell pressure, as shown by the Mohr's circles in Figure 4.13.

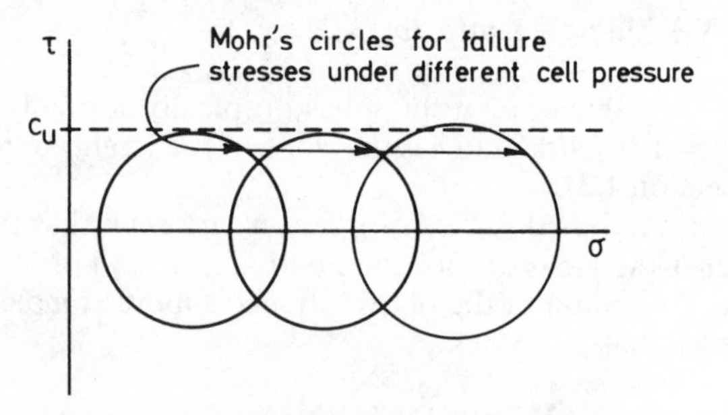

Figure 4.13 Results from quick undrained triaxial tests

Under these conditions the material appears to behave as a Tresca type material, with a failure criterion

$$\frac{\sigma_1'-\sigma_3'}{2} = \frac{\sigma_1-\sigma_3}{2} = \tau_{max} = c_u \tag{4.31}$$

The parameter c_u is known as the undrained shear strength. The 'unconsolidated undrained' tests used to determine it are relatively quick, simple and therefore cheap to perform. The simplest version is the unconfined compression test in which no confining pressure is used. The failure criterion applies in terms of total as well as effective stresses and therefore permits analyses to be performed without knowledge of pore pressures in problems where similar conditions apply to soil elements in the ground. The problems involve loading or unloading over a time period which allows a negligible amount of drainage to occur.

Such 'short term' analyses are usually applicable only to fine grained soils. Note that there is no contradiction between this behaviour and the frictional failures invoked by Mohr-Coulomb and the critical state model; in an undrained situation the soil develops just sufficient pore pressures to allow a frictional failure to occur under effective stresses in accordance with Equation (4.22) when the maximum shear stress is equal to c_u. For the total-stress analysis the pore pressures need not be known, but some idea of their magnitude may be required to allow an accurate prediction of soil behaviour in the longer term as drainage occurs, as discussed in the next section.

The undrained shear strength depends only on the initial void ratio of the soil, irrespective of whether it is normally consolidated or overconsolidated and independent of the stress path to failure. It increases exponentially with decreasing void ratio, and hence for normally consolidated soils increases linearly with the vertical effective stress. The rate of increase varies with the plasticity of the soil, an empirical relationship due to Skempton giving

$$c_u/\sigma'_v = 0.11 + 0.0037.\mathrm{PI} \qquad (4.32)$$

4.9 Pore pressure parameters

From the discussion in section 4.6 it is clear that the pore pressures generated during loading of soil elements may be very different depending on their overconsolidation ratio, even for soil elements with similar initial void ratios and following the same simple total stress path. In the field, soil elements will have different void ratios and will generally be subjected to different and more complex stress paths, and the pore pressure response will vary accordingly.

Conventionally the pore pressure changes are described in terms of parameters A and B where

$$\Delta u = B[\Delta\sigma_3 + A(\Delta\sigma_1 - \Delta\sigma_3)] \qquad (4.33)$$

B and A may be determined from the first and third stages of a consolidated undrained triaxial test in which pore pressures are measured. B should be unity for a saturated soil; it will be less than unity for an unsaturated soil but such soils are outside the scope of this book. A gives an indication of the pore pressures generated during shear; it is clear from Figures 4.10 and 4.11 that it is not constant even during the course of a single test, but is nevertheless useful in giving a qualitative indication of behaviour. Typical values lie between +1.0 for a normally consolidated clay to −0.5 for a heavily overconsolidated one.

Note that overconsolidated clays under increasing stresses, and many soils under decreasing stresses (such as occur below an excavation, or near to a cut slope), may generate negative pore pressure changes under short term loading conditions. For pore pressure equilibrium to be re-established in the long term water will be drawn into the soil, its void ratio will increase and its strength will fall. In fine grained soils such a process may lead to failures occurring many years later. When pore pressures increase under short term loading the opposite occurs and the soil strength will increase with time, giving an increased factor of safety.

4.10 Elastic parameters

In many cases it is satisfactory to use elastic analyses to determine stresses occurring in the ground due to applied loading. For cases in which the resulting deformations are small and the soil is not close to failure it is also satisfactory to determine the deformations from elastic analyses. Values for the elastic parameters E and ν, Young's modulus and Poisson's ratio respectively are then required. These may be obtained from the appropriate (usually initial) part of the q/ε_1 and $\varepsilon_v/\varepsilon_1$ curves, since

$$E' = \frac{\Delta\sigma_1'}{\Delta\varepsilon_1} = \frac{\Delta q}{\Delta\varepsilon_1}\ (\sigma_3' \text{ constant}) \tag{4.34}$$

and

$$\frac{\Delta\epsilon_v}{\Delta\epsilon_1} = 1 - 2\nu', \text{ or } \nu' = \tfrac{1}{2}\left(1 - \frac{\Delta\epsilon_v}{\Delta\epsilon_1}\right) \tag{4.35}$$

The above parameters would be determined from a drained test in which volume change is allowed. E', ν' are therefore effective stress parameters appropriate to long term conditions.

For short term conditions equivalent total stress parameters E_u, ν_u may be obtained from an undrained test. In this case $\Delta\varepsilon_v$ is zero and ν_u is therefore 0.5.

Note that from the theory of elasticity

$$\frac{E_u}{2(1+\nu_u)} = G_u = G' = \frac{E'}{2(1+\nu')} \tag{4.36}$$

where G is the shear modulus which is not affected by pore pressures since the water offers no resistance to shearing.

In practice elastic parameters are very much more sensitive to disturbance of the soil than the strength parameters c_u and ϕ', and great care is required to obtain realistic values from laboratory tests. If possible it may be better to obtain values from large scale field loading tests or by back analysis of the behaviour of full scale constructions.

4.11 Field tests

Many types of soil, including most coarse grained ones, very soft clays and heavily fissured hard clays, are difficult or impossible to recover for laboratory testing without excessive sample disturbance. For these soils a number of types of test have been developed in which the soil is loaded or deformed while still in the ground, either by performing the test in a borehole or by driving a probe into the ground.

Examples of these are the standard penetration test, in which the density of a granular soil is estimated from the number of blows of a standard hammer needed to drive a sampling tube a standard distance; the Dutch cone, in which the force needed to push a cone-shaped probe into the ground is measured and related to the density or strength of the soil; the pressuremeter, in which a flexible cylinder is inflated against the soil and the pressure-deformation curve used to determine soil parameters; and the vane test, in which a metal cruciform of the shape shown in Figure 4.14 is pushed into the ground and then rotated to shear the soil around it. New instruments, and refinements to existing ones, are continually being introduced.

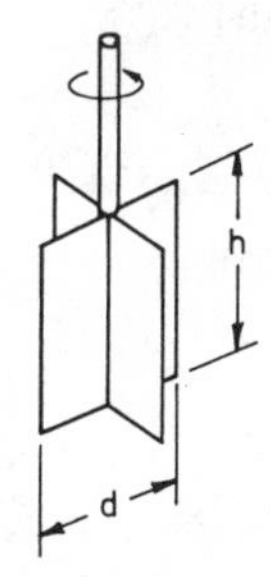

Figure 4.14 The shear vane

Analysis of most of these tests is at present semi-empirical and sometimes related only to particular methods of design. However the vane test, when used in fine grained impermeable soil, may be interpreted directly in terms of the undrained strength and is therefore frequently used in soft clays. It is assumed that shearing takes place on the cylindrical surface immediately surrounding the vane; then the torque T needed to rotate the vane is given by

$$T = \pi c_u(d^2h/2 + d^3/6) \quad (4.37)$$

WORKED EXAMPLES

Example 4.1 Calculation of settlement-time relationship

Write a program to generate the theoretical relationship between U and T_v shown as curve 1 in Figure 4.7, by solution of Equation (4.13).

Terzaghi's theory of one dimensional consolidation may be used to obtain a theoretical relationship between the non-dimensional settlement U and the non-dimensional time factor T_v for the oedometer test. The calculation makes use of Fourier series, and the final result can only be expressed as the sum of an infinite number of terms:

$$U = 1 - \sum_{m=1}^{\infty} \frac{2}{M^2} \exp(-M^2 T_v) \tag{4.38}$$

where $M = \frac{1}{2}\pi(2m - 1)$

It is usual to calculate values of U by truncating the infinite series after a certain number of terms, and the purpose of Example 4.1 is to carry out this calculation. The number of terms in the series is input at the beginning of the program, so that several calculations may be made using different numbers of terms. In this way the minimum number for a reasonably accurate calculation may be found (it is surprisingly small).

```
LIST

10      REM EXAMPLE 4.1. CALCULATION OF SETTLEMENT USING TERZAGHI THEORY
20      REM
30      PRINT "ENTER NUMBER OF TERMS IN SUMMATION EQUATION:"
40      INPUT N
50      PRINT " "
60      PRINT "TV","U"
70      PRINT " "
80      PRINT 0,0
90      REM
100     REM MAIN LOOP FOR CALCULATION OF U
110     REM
120     FOR I=1 TO 12
130     TV = 0.1 * I
140     REM
150     REM SUMMATION OF TERMS
160     REM
170     U=1
180     FOR J=0 TO N
190     M = 3.14159 * 0.5 * (2 * J - 1)
200     U = U - 2 * EXP(- M * M * TV) / (M * M)
210     NEXT J
220     PRINT TV,U
230     NEXT I
240     STOP
250     END

READY

RUN

ENTER NUMBER OF TERMS IN SUMMATION EQUATION:
? 3

TV              U

 0              0
 .1             .356822
 .2             .504087
```

```
.3            .613235
.4            .697881
.5            .763949
.6            .815564
.7            .855892
.8            .887402
.9            .912023
1             .931259
1.1           .94629
1.2           .958034
```

Program notes

(1) The program is self explanatory, and begins with input statements and the printing of headings. The main part of the program consists of two loops. The main loop (120 to 230) covers the range of T_v values from 0.1 to 1.2, and the inner loop (180 to 210) carries out the summation of the terms in the equation. Note the use of the 'temporary' variable M which is used to make the program both clearer and faster.
(2) This program illustrates an important application of computing in carrying out repetitive theoretical calculations where numerical values are required.

Example 4.2 Calculation of voids ratio from oedometer tests

Write a program to analyse the results of an oedometer test.

The main purpose of an oedometer test is to provide a plot of voids ratio against vertical effective stress. The readings obtained from the oedometer are, however, the movement of a dial gauge and the weight on the loading hanger. Example 4.2 shows how a program can be used to carry out standard calculations to process the data and give the required quantities of voids ratio and vertical effective stress.

```
LIST

10       REM EXAMPLE 4.2. CALCULATE SIGMA-V AND E FROM CONSOLIDATION TEST
20       REM
30       REM DIAMETER OF CONSOLIDATION RING (MM)
40       DATA 75
50       REM HEIGHT OF CONSOLIDATION RING (MM)
60       DATA 15
70       REM LEVER ARM FOR WEIGHTS
80       DATA 7
90       REM
100      DIM W(20),R(20)
110      READ D
120      READ HO
130      READ L
140      A = 1E-6 * D * D * 3.14159 / 4
150      PRINT "ENTER INITIAL VOIDS RATIO:"
160      INPUT EO
170      PRINT "ENTER INITIAL DIAL GAUGE READING (ZERO LOAD) (MM):"
180      INPUT RO
190      N = 0
200      PRINT "ENTER WEIGHT ON HANGER, DIAL GAUGE (0, 0 TO END) (KGF, MM):"
```

```
210        N = N+1
220        INPUT W(N),R(N)
230        IF W(N)<>0 GOTO 210
240        N = N-1
250        PRINT " "
260        PRINT "VOIDS RATIO","SIG-V","LN(SIG-V)"
270        PRINT " ","KN/M2"
280        PRINT " "
290        FOR I=1 TO N
300        P = W(I) * L * 9.81 * 0.001
310        H = HO + RO - R(I)
320        SV = P / A
330        E = ((1 + EO) * H / HO) - 1
340        PRINT E,SV,LOG(SV)
350        NEXT I
360        STOP
370        END

READY

RUN

ENTER INITIAL VOIDS RATIO:
? .975
ENTER INITIAL DIAL GAUGE READING (ZERO LOAD) (MM):
? 1.698
ENTER WEIGHT ON HANGER, DIAL GAUGE (0, 0 TO END) (KGF, MM):
? 2, 1.754
? 5, 1.965
? 10, 2.577
? 20, 3.168
? 40, 3.789
? 80, 4.491
? 20, 4.151
? 2, 3.67
? 0, 0

VOIDS RATIO    SIG-V          LN(SIG-V)
               KN/M2

 .967627        31.0874        3.4368
 .939845        77.7186        4.35309
 .859265        155.437        5.04624
 .78145         310.874        5.73939
 .699685        621.749        6.43254
 .607255        1243.5         7.12568
 .652022        310.874        5.73939
 .715353        31.0874        3.4368
```

Program notes

(1) The lines 30 to 80 make use of the DATA statement to provide the values of certain quantities which will usually remain constant — in this case the dimensions in millimetres of the oedometer and the lever arm of the loading arrangement. These statements need only be changed if a different oedometer is used. The data statements are read using the READ statements at lines 110 to 130. The variable data for each test is then read in — again using a system where the data is terminated by inputting a special number.

(2) As well as calculating the voids ratio and vertical effective stress, the logarithm of the stress is also output since this is often used for plotting. It is always useful to print out any quantity that may be of use later — there is no point in carrying out further processing by hand if the computer could have done it in the first place.

Example 4.3 Analysis of drained triaxial tests

Write a program to analyse the results of a triaxial compression test.

The triaxial test is one of the commonest and most useful tests in the soil mechanics laboratory. The analysis of the test involves a tedious repetitive calculation, and it is an ideal problem for computer analysis.

Example 4.3 is a program for the analysis of a drained triaxial test.

```
LIST

10      REM EXAMPLE 4.3. ANALYSIS OF DRAINED TRIAXIAL TEST
20      REM
30      DIM AL(50),AD(50),VC(50),SA(50),EA(50),ER(50)
40      DIM P(50),Q(50),ES(50),EV(50),ET(50)
50      PRINT "DRAINED TRIAXIAL TEST ANALYSIS"
60      PRINT "ENTER INITIAL DIAMETER (MM):"
70      INPUT DO
80      PRINT "ENTER INITIAL HEIGHT (MM):"
90      INPUT HO
100     PRINT "ENTER CELL PRESSURE (KN/M2):"
110     INPUT C
120     PRINT "ENTER BACK PRESSURE (KN/M2):"
130     INPUT B
140     PRINT "ENTER RAM LOAD BEFORE CONTACT WITH SAMPLE (N):"
150     INPUT LO
160     K = 0
170     PRINT "ENTER AXIAL LOAD (N), AXIAL DEF. (MM), VOL. CHANGE (CC)"
180     PRINT "(0, 0, 0 TO END):"
190     K = K + 1
200     INPUT AL(K),AD(K),VC(K)
210     IF AL(K)<>0 GOTO 190
220     K = K - 1
230     SR = C - B
240     AO = 3.14159 * DO * DO / 4E6
250     VO = AO * HO * 0.001
260     FOR I=1 TO K
270     H = (HO - AD(I)) * 0.001
280     V = VO - VC(I) * 1E-6
290     A = V / H
300     Q(I) = (AL(I) - LO) * 0.001 / A
310     SA(I) = SR + Q(I)
320     P(I) = (SA(I) + SR + SR) / 3
330     EA(I) = 100 * AD(I) / HO
340     EV(I) = 0.0001 * VC(I) / VO
350     ER(I) = (V(I) - EA(I)) / 2
360     ES(I) = (EA(I) - ER(I)) / 1.5
370     ET(I) = Q(I)/P(I)
380     NEXT I
390     PRINT " "
400     PRINT "SIGMA-A","SIGMA-R","EPSILON-A","EPSILON-R"
410     PRINT "(KN/M2)","(KN/M2)","(PERCENT)","(PERCENT)"
420     PRINT " "
430     FOR I=1 TO K
440     PRINT SA(I),SR,EA(I),ER(I)
450     NEXT I
460     PRINT " "
470     PRINT "P","Q","ETA","EPS-V","EPS-S"
480     PRINT "(KN/M2)","(KN/M2)"," ","(PERCENT)","(PERCENT)"
490     PRINT " "
500     FOR I=1 TO K
510     PRINT P(I),Q(I),ET(I),EV(I),ES(I)
520     NEXT I
530     STOP
540     END

READY
```

```
RUN

DRAINED TRIAXIAL TEST ANALYSIS
ENTER INITIAL DIAMETER (MM):
? 37.8
ENTER INITIAL HEIGHT (MM):
? 76
ENTER CELL PRESSURE (KN/M2):
? 258
ENTER BACK PRESSURE (KN/M2):
? 103
ENTER RAM LOAD BEFORE CONTACT WITH SAMPLE (N):
? 2.7
ENTER AXIAL LOAD (N), AXIAL DEF. (MM), VOL. CHANGE (CC)
(0, 0, 0 TO END):
? 50.5, 0.37, 0.63
? 98.1, 0.96, 1.24
? 127, 1.51, 1.78
? 151.5, 2.1, 1.96
? 167.8, 3.21, 2.1
? 179.2, 4.6, 2.4
? 184.3, 5.9, 2.67
? 186.7, 7.87, 2.78
? 187.4, 9.77, 2.81
? 184.2, 11.2, 2.82
? 0, 0, 0

SIGMA-A        SIGMA-R        EPSILON-A      EPSILON-R
(KN/M2)        (KN/M2)        (PERCENT)      (PERCENT)

 197.703        155            .486842        .125917
 240.176        155            1.26316        .953717E-01
 265.877        155            1.98684        .501047E-01
 286.965        155            2.76316       -.232528
 299.464        155            4.22368       -.880716
 307.038        155            6.05263       -1.61931
 309.085        155            7.76316       -2.31629
 306.936        155            10.3553       -3.54786
 303.315        155            12.8553       -4.78027
 297.616        155            14.7368       -5.71519

P              Q              ETA            EPS-V          EPS-S
(KN/M2)        (KN/M2)                       (PERCENT)      (PERCENT)

 169.234        42.7027        .252329        .738676        .240617
 183.392        85.1756        .464446        1.4539         .778524
 191.959        110.877        .577609        2.08705        1.29116
 198.988        131.965        .663178        2.2981         1.99712
 203.155        144.464        .711103        2.46225        3.40293
 205.679        152.038        .7392          2.814          5.11463
 206.362        154.085        .746675        3.13058        6.71963
 205.645        151.936        .738826        3.25955        9.26875
 204.438        148.315        .725475        3.29473        11.757
 202.539        142.616        .704141        3.30645        13.6347
```

Program notes

(1) The first part of the program (up to line 220) represents the now familiar pattern of DIM statements, input of preliminary variables (cell pressure etc.) and input of the main series of data using a special value (zero) to terminate the data.
(2) The radial effective stress SR is calculated, and also the initial cross-sectional area and volume. Note that various factors of powers of ten appear in the calculation. This is because of the mixture of units used for the input calculation, and is a fairly

common state of affairs. Great care must be taken in ensuring that the correct units are used.
(3) All the stresses and strains are calculated in one loop (260 to 380) using the assumption that the sample remains cylindrical.
(4) The output must then be in two stages — first the principal stresses and strains and then the stresses and strains in p,q parameters, since the computer screen is not sufficiently wide to print all the variables on a single line.

Example 4.4 'Least squares' analysis of test results

Experimental data in soil mechanics often contains a considerable amount of scatter, and one common problem is that of fitting a straight line through a set of experimental data points. For instance a series of measurements of undrained strengths at different depths may be available and the design engineer may want to fit these by a linear variation of strength with depth. Example 4.4 is a program which uses the least squares method to fit a series of experimental points. The program has been written in general terms, with a series of pairs of values of co-ordinates x,y. The program fits the best straight line $y = y_0 + mx$ through the points, and also gives the root mean square error in the y value. The program could be modified for a more specialized application.

The derivation of the best fit straight line is a standard piece of mathematics which will be found in most textbooks (see for instance *Advanced Engineering Mathematics* by E. Kreyszig, John Wiley (1962)) and is not repeated here.

```
LIST

10        REM EXAMPLE 4.4. GENERAL LEAST SQUARES ANALYSIS PROGRAM
20        REM ENTER LIST OF X AND Y VALUES, PROGRAM FITS BEST FIT LINE
30        REM Y = YO + M * X
40        REM
50        DIM X(100),Y(100)
60        PRINT "LEAST SQUARES FIT OF STRAIGHT LINE"
70        PRINT " "
80        PRINT "ENTER PAIRS OF X, Y (0, 0 TO END DATA)"
90        N = 0
100       N = N + 1
110       INPUT X(N),Y(N)
120       IF X(N)<>0 GOTO 100
130       IF Y(N)<>0 GOTO 100
140       N = N - 1
150       SX = 0
160       SY = 0
170       XX = 0
180       XY = 0
190       FOR I = 1 TO N
200       SX = SX + X(I)
210       SY = SY + Y(I)
220       XX = XX + X(I)*X(I)
230       XY = XY + X(I)*Y(I)
240       NEXT I
250       YO = (SY*XX - SX*XY) / (N*XX - SX*SX)
```

```
260      M = (SY - N*YO) / SX
280      E = 0
290      FOR I = 1 TO N
300      E = E + (Y(I) - YO - M*X(I)) ^ 2
310      NEXT I
320      RM = SQR(E / N)
330      PRINT "BEST FIT IS: Y =";YO;" +";M;" * X"
340      PRINT "RMS ERROR IS:";RM
350      STOP
360      END

READY

RUN

LEAST SQUARES FIT OF STRAIGHT LINE

ENTER PAIRS OF X, Y (0, 0 TO END DATA)
? 2.3, 21.4
? 4.5, 31.4
? 4.8, 29.8
? 6.7, 45.0
? 9.0, 50.3
? 12.5, 80.3
? 13.5, 60.0
? 21.0, 94.3
? 22.0, 114.0
? 0, 0
BEST FIT IS: Y = 12.3915   + 4.30921   * X
RMS ERROR IS: 7.11254
```

Program notes

(1) The input to the program follows a very similar pattern to some of the previous programs, again using a pair of special values of x and y (0,0) to terminate the list of input data. This saves the programmer the effort of having to count the number of data points first.

(2) The main part of the least squares analysis requires summation of various terms (x, y, x^2 etc.). This is achieved by setting some variables to zero (lines 150 to 180) and then using a FOR . . . NEXT loop to add up the terms in the summation.

(3) The output from the program is simple. The semicolon separator is used in the print statements to provide a readable and clear form of output.

Example 4.5 Prediction of strains using Cam clay model

Write a program to predict the strains for a clay subjected to a given change in stresses, using the Cam clay theory. The soil properties, M, λ, κ, v and N will be required, and the starting conditions of the sample defined by p', q and p_c.

```
LIST

10      REM EXAMPLE 4.5. CAM-CLAY PREDICTIONS OF STRAINS FOR TRIAXIAL TESTS
20      REM
30      PRINT "ENTER SOIL PROPERTY VALUES"
40      PRINT "MU"
50      INPUT MU
60      PRINT "LAMBDA"
70      INPUT L
80      PRINT "KAPPA"
90      INPUT K
100     PRINT "NU"
110     INPUT NU
120     PRINT "N"
130     INPUT N
140     PRINT "ENTER INITIAL VALUES:"
150     PRINT "P (KN/M2)"
160     INPUT P
170     PRINT "Q (KN/M2)"
180     INPUT Q
190     PRINT "PC (KN/M2)"
200     INPUT PC
210     PRINT "ENTER INCREMENT INFORMATION:"
220     PRINT "INCREMENT IN P (KN/M2)"
230     INPUT DP
240     PRINT "INCREMENT IN Q (KN/M2)"
250     INPUT DQ
260     PRINT "NUMBER OF INCREMENTS"
270     INPUT NI
280     PRINT " "
290     PRINT "P","Q","EPS-V","EPS-S","V"
300     PRINT "KN/M2","KN/M2","PERCENT","PERCENT"
310     PRINT " "
320     EV=0
330     ES=0
340     V = N - L*LOG(PC) - K*LOG(P/PC)
350     PRINT P,Q,0,0,V
360     FOR I=1 TO NI
370     EV = EV + K*DP/(P*V)
380     ES = ES + K*(1+NU)*DQ/(4.5*(1-2*NU)*P*V)
390     I$="ELAS"
400     IF (P * EXP(Q / (MU*P))) < 0.99*PC GOTO 460
410     I$="PLAS"
420     DV = (L - K) * (DP*(MU-Q/P) + DQ) / (MU*P*V)
430     DS = DV / (MU - Q/P)
440     EV = EV + DV
450     ES = ES + DS
460     P = P + DP
470     Q = Q + DQ
480     IF I$="PLAS" THEN PC = P * EXP(Q / (MU*P))
490     V = N - L*LOG(PC) - K*LOG(P/PC)
500     PRINT P,Q,EV*100,ES*100,V,I$
510     IF Q>MU*P THEN PRINT "WARNING Q>MP - RESULTS MAY NOT BE VALID"
520     NEXT I
530     STOP
540     END

READY

RUN

ENTER SOIL PROPERTY VALUES
MU
? 1.08
LAMBDA
? 0.21
KAPPA
? 0.043
```

```
NU
? 0.2
N
? 3.3
ENTER INITIAL VALUES:
P (KN/M2)
? 110
Q (KN/M2)
? 0
PC (KN/M2)
? 150
ENTER INCREMENT INFORMATION:
INCREMENT IN P (KN/M2)
? 5
INCREMENT IN Q (KN/M2)
? 15
NUMBER OF INCREMENTS
? 10
```

P KN/M2	Q KN/M2	EPS-V PERCENT	EPS-S PERCENT	V	
110	0	0	0	2.2611	
115	15	.864421E-01	.115256	2.25919	ELAS
120	30	.169196	.225594	2.25736	ELAS
125	45	1.34171	1.64847	2.23039	PLAS
130	60	2.45044	3.18408	2.20645	PLAS
135	75	3.50072	4.86104	2.18399	PLAS
140	90	4.49785	6.72053	2.16285	PLAS
145	105	5.44647	8.82282	2.14291	PLAS
150	120	6.35068	11.2615	2.12406	PLAS
155	135	7.2141	14.1941	2.1062	PLAS
160	150	8.03994	17.9177	2.08925	PLAS

Program notes

(1) Lines 10 to 270 input the soil properties and starting conditions. The strains ES and EV are then initialized to zero and the initial specific volume v calculated.
(2) At lines 370 and 380 the increment in strain is calculated assuming the behaviour to be elastic. Line 390 sets an indicator to the text string "ELAS".
(3) At line 400 the next few lines are skipped if the soil is within the yield locus. (If it is within 1% of it, it is assumed that the current increment will reach the locus, so the plastic equations are used).
(4) Lines 410 to 450 add on the plastic strain if necessary, and 460 and 470 update the stress values. Line 480 updates the p_c value only if the indicator has been changed to "PLAS" at 410. Finally line 490 updates the specific volume value.
(5) Line 510 prints a warning in the case of $q > Mp'$. A treatment of this case is beyond the scope of this book.
(6) The example shows a calculation for a drained triaxial test on a silty clay. The clay has first been isotropically consolidated to $p' = 130$ kN/m^2 (resulting in $p_c = 130$ kN/m^2) and then allowed to swell to $p' = 100$ kN/m^2. It is then sheared with $\sigma_3' =$ constant, so that $dq = 3\ dp'$. Note that the model predicts that the clay compresses as it is sheared.

PROBLEMS

(4.1) If your computer has any graphics facilities modify the program in Example 4.2 to plot a curve of U against T_v.

(4.2) In a laboratory there are three consolidation devices, codenamed CON1, CON2 and CON3, each using different sample sizes. Modify the first part of the program in Example 4.1 so that the operator enters the codename for the device and the appropriate dimensions are read from a table which is given as part of the program. (It may be useful to use a character variable, say C$, and statements of the form IF C$ = "CON1" THEN. . .).

(4.3) Write a program for the analysis of undrained triaxial tests. The output should be similar to that in Example 4.3, except that the volume change is not of interest and both total and effective stresses, as well as the pore pressure, should be output. The input data should be assumed to be in the form of axial load, axial deformation and pore pressure for each data point. The cell pressure is kept constant.

(4.4) The program in Example 4.4 is very simple and gives only the minimum of output. Modify it in two ways:

(a) include a check that at least two data points have been entered (so that a straight line can be fitted) and a check that the maximum number of data points (100) is not exceeded.

(b) add to the output some additional information which should include (i) the total number of data points (ii) a table showing the input x and y values of each point together with the predicted **y** value from the best fit line and the resulting error (iii) (more difficult) the maximum error as well as the root mean square value, and the number of data points for which the maximum error occurs.

(4.5) The values of strains calculated when using the Cam clay theory depend on the size of the increments of stress used. A large number of increments are required to obtain accurate answers, but are expensive in computer time. Use the program in Example 4.5 to assess what is a suitable increment size for a variety of different stress paths.

(4.6) The program in Example 4.5 is only suitable for calculating the strains which result from changes in stress. Many problems involve control of strain, and it is the stresses which are the unknowns to be calculated. For instance, in an undrained triaxial test $d\varepsilon_v$ is equal to zero and $d\varepsilon_1$ (and hence $d\varepsilon_s$) is increased.

Equations (4.23) to (4.30) may be rearranged to express the stress-strain relations in a matrix form:

$$\begin{bmatrix} d\varepsilon_v \\ d\varepsilon_s \end{bmatrix} = \begin{bmatrix} \kappa/p'v & 0 \\ 0 & 2\kappa(1+\nu)/9p'v(1-2\nu) \end{bmatrix} \begin{bmatrix} dp' \\ dq \end{bmatrix} \tag{4.39}$$

for elastic behaviour and:

$$\begin{bmatrix} d\varepsilon_v \\ \\ d\varepsilon_s \end{bmatrix} = \tag{4.40}$$

$$\begin{bmatrix} \dfrac{\kappa}{p'v} + \dfrac{(\lambda-\kappa)}{Mp'v}(M-q/p) & \dfrac{(\lambda-\kappa)}{Mp'v} \\ \dfrac{(\lambda-\kappa)}{Mp'v} & \dfrac{2\kappa(1+\nu)}{9p'v(1-2\nu)} + \dfrac{\lambda-\kappa}{Mp'v(M-q/p')} \end{bmatrix} \begin{bmatrix} dp' \\ \\ dq \end{bmatrix}$$

for elastic-plastic behaviour. By inverting these matrices it is possible to express dp' and dq in terms of $d\varepsilon_v$ and $d\varepsilon_s$.

Write a program similar to Example program 4.5 but designed to calculate stresses given increments of strain. It will be most convenient to calculate the terms in the above matrices and then invert the matrices in the program (do not attempt to do this algebraically).

The inverse of a 2 × 2 matrix is given by:

$$\begin{bmatrix} a & b \\ c & d \end{bmatrix}^{-1} = \begin{bmatrix} d/D & -b/D \\ -c/D & a/D \end{bmatrix} \tag{4.41}$$

where $D = ad - bc$.

Use the program to predict the stress path for an undrained triaxial test on a material with the same properties as used in Example 4.5. The initial conditions are $p' = 100$ kN/m^2, $q = 0$ kN/m^2 and $p_c = 130$ kN/m^2. Consider 20 increments of ε_s each of 0.5 per cent.

Chapter 5

Earth pressures

ESSENTIAL THEORY

5.1 Plasticity theorems and soil models

For the solution of many problems the soil is idealized as a rigid-plastic material; elastic deformations prior to yield are ignored, as is the increase in yield stress with continuing strain (work-hardening). The condition for yielding to occur is taken to be the Mohr-Coulomb criterion. Methods of analysis may be derived for soils that have both cohesion and internal friction, but in most cases the soil may be considered as purely cohesive ($\phi' = 0$) or purely frictional ($c = 0$) as discussed in Chapter 4. The former is appropriate for total stress analyses of soil under undrained conditions, usually applicable to the short term behaviour of fine grained soils. The latter is appropriate for effective stress analyses of soil under fully drained conditions, applicable to coarse grained soils in almost all cases and fine grained soils in the long term.

The collapse of soil structures may then be considered in terms of the theorems of plasticity, which state that for a perfectly plastic material subjected to loads which are increased in proportion:
(i) there exists a unique set of collapse loads
(ii) if a stress distribution within the material can be found which is in equilibrium with the loads and does not violate the yield criterion, then the loads will be less than or equal to the collapse loads (the lower bound theorem)
(iii) if a compatible failure mechanism can be found which allows the external loads to do work greater than or equal to the work dissipated internally, then the loads will be greater than or equal to the collapse loads (the upper bound theorem)

For a detailed explanation of these theorems see *Engineering Plasticity* by C. R. Calladine (Pergamon).

Calculations of both upper and lower bound types are used in soil mechanics; obviously if both types of calculation give the same result for a particular problem then the actual collapse loads have been found. In other cases it may be possible to establish a solution within reasonably narrow bounds. The theorems are theoretically strictly applicable to undrained conditions only, but are found in practice to give a reasonable description of behaviour for drained conditions.

In many cases calculations of a type known as limit-equilibrium are performed. For these a mode of failure is assumed and the equilibrium of forces on a soil mass considered just as failure is about to occur. Such a calculation gives neither strictly upper nor lower bound solutions, but provided the failure mode is reasonable is found to give a good approximation to the actual collapse loads.

Examples of each type of calculation follow; for a much more detailed discussion see Reference 8.

5.2 Rankine-Bell stress states, active and passive pressure

The behaviour of a soil mass which is in equilibrium throughout and has everywhere just reached yield was first considered by Rankine and the theory later extended by Bell. The approach is particularly useful for the calculation of earth pressures against vertical retaining walls. It is assumed that the vertical stress (σ_v) on a soil element close to the wall is a principal stress determined from the weight of material above the element plus any surcharge on the surface. The horizontal stress may have one of two limiting values (see Figure 5.1), a minimum 'active' value when the wall is moving away from the soil and a maximum 'passive' value when the wall is moving towards the soil. In either case the soil is just yielding on planes (slip lines) at angles of $(45° - \phi'/2)$ to the direction of maximum principal stress (see Figure 5.1). The active and passive pressures are given by

$$\sigma_a = K_a\sigma_v - 2c\sqrt{K_a} \tag{5.1}$$

$$\sigma_p = K_p\sigma_v + 2c\sqrt{K_p} \tag{5.2}$$

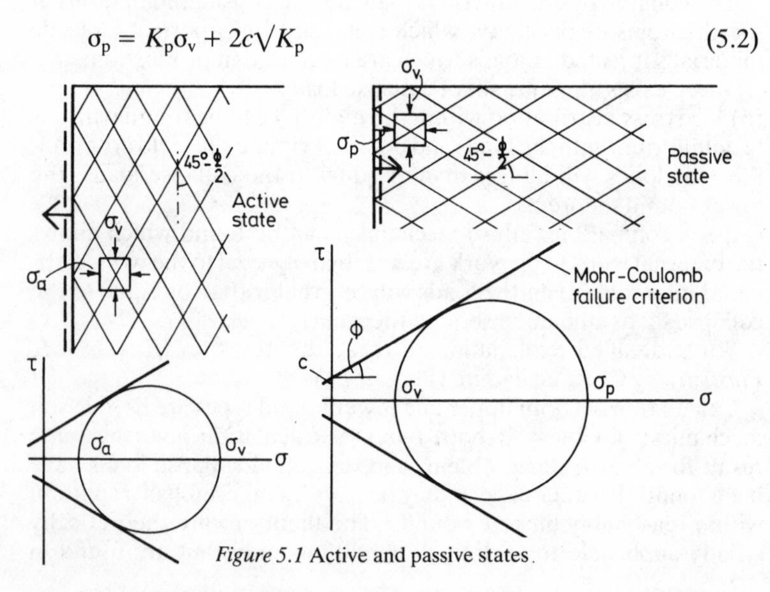

Figure 5.1 Active and passive states.

where K_a and K_p are the coefficients of active and passive pressure respectively, given by

$$K_a = \frac{1}{K_p} = \frac{1 - \sin\phi}{1 + \sin\phi} = \tan^2(45° - \frac{\phi}{2}) \tag{5.3}$$

Note that in layered soils the values of K_a and K_p will be different in the different soils. Equations (5.1) and (5.2) may be used either for short term total-stress calculations under undrained conditions, or for long term effective-stress calculations under drained conditions. In the latter case any water pressures in the soil must be added to the effective earth pressures to give the total pressures on the wall. In the former case $K_a = K_p = 1$ and Equation (5.1) predicts a negative value of σ_a for depths for which σ_v is less than $2c_u$. In practice the soil will tend to fail in tension to such a depth, forming a tension crack which under wet conditions may fill with water. The tension crack may extend to a depth of $2c_u/\gamma$.

Earth pressures on retaining walls, particularly in the active state, are commonly associated with decreasing mean stress levels. Soils will therefore tend to expand and soften and conditions in the long term will be more critical than in the short term.

Solutions using Rankine-Bell stress states are lower bound solutions and give an overestimate of active pressures and underestimate of passive pressures. If the ground surface behind the wall is sloping, or there are shear stresses on the wall due to friction or adhesion between the wall and the soil, the vertical stress will in general no longer be a principal stress and the simple theory will not apply. Slip line fields may be determined and analysed for such cases but in practice a limit-equilibrium approach as discussed in section 5.4 is normally used.

5.3 Braced cuts in soft clay

Temporary excavations in soft cohesive soils are frequently required. It is normal to support the side of the excavation with sheeting supported by props, as shown in Figure 5.2. The main problem is to determine the total force to be resisted by the props. If it is assumed that sufficient movement can take place for active conditions to be developed, and the soil has uniform undrained shear strength c_u and unit weight γ, the total active force on the wall determined from Equation (5.1) will be

$$P_{a1} = \frac{\gamma H^2}{2} - 2c_u H \tag{5.4}$$

However it may be possible for the soil to fail in the way shown in Figure 5.2, slumping down behind the wall and heaving up within

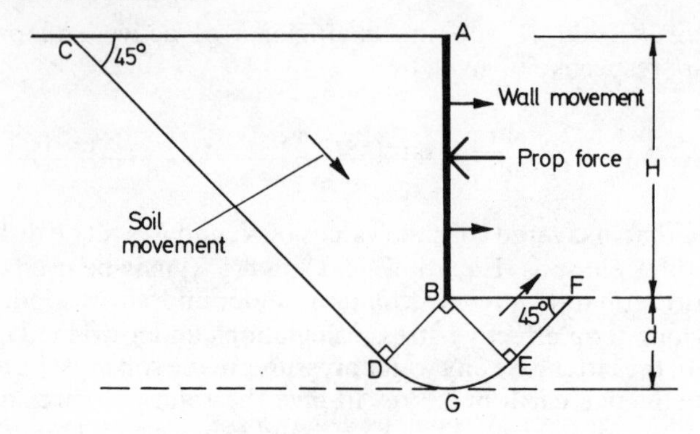

Figure 5.2 Failure of a braced cut in soft clay

the excavation. An upper bound calculation due to Henkel is based on the compatible system of deformations shown, work being done by the self-weight of the blocks of soil and dissipated by the shear forces at the boundaries between the blocks and within the continuously deforming zone BDGE. By equating the work, it can be shown that

$$P_{a2} = \frac{\gamma H^2}{2} - 2c_u H + \sqrt{2} d\gamma H - \sqrt{2} d(\pi + 2)c_u \tag{5.5}$$

For a cut of limited width B, the depth of failure d will be limited to $B/\sqrt{2}$, otherwise it may be limited by a lower stratum of stronger soil. c_u in the second term applies to the soil above the excavation level, in the fourth term to the soil below the excavation level, and different values may be used if appropriate. P_{a2} will be greater than P_{a1} if the third term exceeds the fourth term in Equation (5.5); this is then clearly the more critical mode of failure.

5.4 Coulomb wedge analyses

The oldest type of analysis in soil mechanics is a limit-equilibrium analysis due to Coulomb which was first published in 1773. It is still much used today, for instance to calculate the thrust on a retaining wall by assuming that the soil behind the wall will fail by sliding as a rigid block of wedge shape as shown in Figure 5.3. In general the forces shown may be present, the only unknowns being the magnitudes of the forces Q and R. W is the weight of the wedge of soil, and δ is the angle of wall friction, whose direction depends on the direction of relative sliding of the wall and adjacent soil, and whose magnitude must be less than ϕ' and depends on the roughness of the wall surface. c_a is similarly the cohesion (or adhesion) term for sliding of cohesive soil over the wall surface. By considering the equilibrium of the wedge the unknown forces may be determined. The result will depend on the inclination selected for the

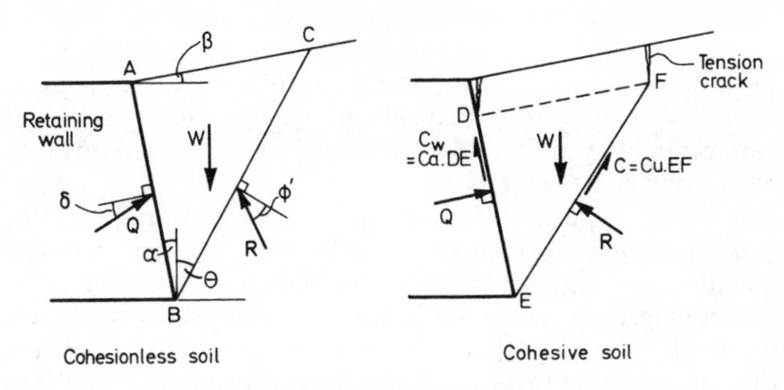

Figure 5.3 Coulomb wedges

failure plane in the soil, and a number of trial calculations using different inclinations must be made to find the most critical value. Graphical methods are often used, but such a procedure is clearly ideal for repeated calculation by computer (see Examples 5.2 and 5.3). The assumption of a plane failure surface is in general an approximation but gives sufficiently accurate results under most conditions of active failure.

Passive forces may be determined in a similar way, with the wedge now moving upwards along the failure surface. The plane surface only gives reasonably accurate answers if the wall friction or adhesion is small, otherwise the failure surface must be represented by a logarithmic spiral curve or a combination of a circular arc and straight lines.

For the particular case of horizontal ground surface, vertical wall and $\delta = c_a = 0$, the Coulomb wedge analysis gives the same results as does Rankine's method, except that the wedge analysis gives no information on the distribution of pressure on the wall, only the resultant force. Under these conditions the most critical failure plane is parallel to one of the sets of slip lines in the Rankine method.

5.5 Wall movements

All the above calculations assume that the wall moves in such a way as to develop fully active or passive conditions. Real soil requires finite strains to reach yield, and wall movements may not be sufficient to cause such strains. If the wall is totally unyielding no strains will develop in the soil and the effective pressures against the wall will be 'at rest' pressures given by

$$\sigma'_w = K_0 \sigma'_v \tag{5.6}$$

For normally consolidated clays and cohesionless materials it is found that K_0 is given approximately by

$$K_0 = 1 - \sin \phi' \tag{5.7}$$

σ'_w is then smaller than σ'_v (typically $\sigma'_w \approx 0.5\sigma'_v$). For overconsolidated clays or granular fill compacted against a wall K_0 may be approximately unity, while for very heavily overconsolidated clays K_0 may be as great as 3.

Typical stress paths to failure for an element of soil under active and passive conditions are shown in Figure 5.4(a) and corresponding load-deflection curves for a wall in Figure 5.4(b). Very small deflections are generally sufficient to develop fully active conditions, while relatively large deflections may be necessary to develop fully passive conditions. For the wall movements that can be tolerated in practice, passive pressures may actually be much less than calculated by the methods given above.

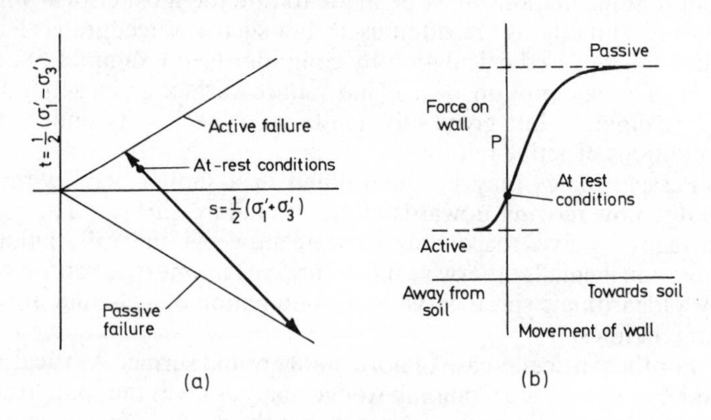

Figure 5.4 Failure under active and passive conditions

The pattern of strains and hence mode of failure in the soil is clearly related to the way in which the wall moves, as shown in Figure 5.5. Most walls move in a combination of modes (a) and (b), which are well modelled by the Rankine and Coulomb methods, but walls whose movements are restricted by anchors or props may induce failures in mode (c).

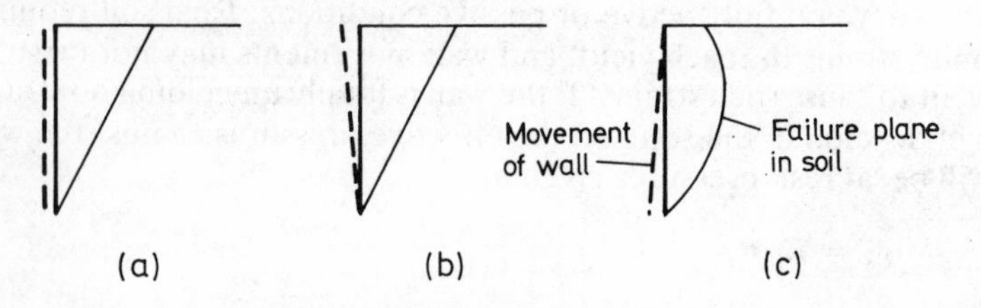

Figure 5.5 Typical wall movements

5.6 Design of retaining walls

There are several different structural forms of retaining wall, and a number of ways in which they can fail; these are discussed in standard textbooks on soil mechanics and geotechnical engineering, and a typical case is covered in Example 5.4. In all cases there is a need to introduce a safety or load factor into the calculation to take account of uncertainties and approximation in the derivation of soil parameters, application of models of soil behaviour and the methods of analysis used. The best way of doing this is a matter of continuing debate, but the conventional method has been to use a safety factor on the available strength of the soil. For retaining walls the 'disturbing' forces — applied loads and active pressures for example — are calculated as accurately as possible and the actual values used. The factor of safety is then defined as the ratio of the total available 'restoring' forces — prop loads and passive soil pressures for example — calculated for material that is just starting to yield, to the actual forces calculated as necessary just to maintain stability.

WORKED EXAMPLES

Example 5.1 Calculation of lateral stresses using Rankine's method

Write a program to determine horizontal active or passive stresses behind a vertical wall retaining layers of different cohesionless soils, using Rankine's theory. Values of K_a and K_p may be determined for each layer from Equation (5.3) and then the horizontal effective stress obtained by multiplying σ_v' by K_a or K_p in accordance with Equations (5.1) and (5.2).

```
LIST

10      REM EXAMPLE 5.1. PROGRAM TO CALCULATE TOTAL AND EFFECTIVE STRESSES
20      REM FOR ACTIVE OR PASSIVE CASES
30      REM
40      DIM G(20),Z(20),P(20)
50      PRINT "ENTER GROUND WATER LEVEL (M BELOW SURFACE):"
60      INPUT W
70      PRINT "ENTER 0 FOR ACTIVE 1 FOR PASSIVE:"
80      INPUT AP
90      K=0
100     PRINT "ENTER DEPTH, BULK UNIT WEIGHT, PHI (M,KN/M3,DEG) (0,0,0 TO END):"
110     K=K+1
120     INPUT Z(K),G(K),P(K)
130     IF Z(K)<>0 GOTO 110
140     K=K-1
150     PRINT " "
160     PRINT "Z","SIG-V(TOT)","SIG-V(EFF)","SIG-H(TOT)","SIG-H(EFF)"
170     PRINT "M","KN/M2","KN/M2","KN/M2","KN/M2"
180     PRINT " "
190     PRINT 0,0,0,0,0
200     TV = 0
210     ZL = 0
220     FOR I=1 TO K
230     TV = TV + (Z(I) - ZL) * G(I)
```

```
240        ZL = Z(I)
250        U = 9.81 * (Z(I) - W)
260        IF U<0 THEN U = 0
270        EV = TV - U
280        S = SIN(P(I)*3.14159/180)
290        KA = (1 - S) / (1 + S)
300        EH = EV * KA
310        IF AP=1 THEN EH = EV / KA
320        TH = EH + U
330        PRINT Z(I),TV,EV,TH,EH
340        NEXT I
350        STOP
360        END

READY

RUN

ENTER GROUND WATER LEVEL (M BELOW SURFACE):
? 3.2
ENTER 0 FOR ACTIVE 1 FOR PASSIVE:
? 0
ENTER DEPTH, BULK UNIT WEIGHT, PHI (M,KN/M3,DEG) (0,0,0 TO END):
? 2, 14.75, 25.5
? 4.3, 17.6, 31
? 5.9, 17.2, 31
? 9.7, 19.83, 31
? 15.6, 18.9, 38.8
? 0, 0, 0

Z              SIG-V(TOT)     SIG-V(EFF)     SIG-H(TOT)     SIG-H(EFF)
M              KN/M2          KN/M2          KN/M2          KN/M2

 0             0              0              0              0
 2             29.5           29.5           11.744         11.744
 4.3           69.98          59.189         29.7374        18.9464
 5.9           97.5           71.013         49.2182        22.7312
 9.7           172.854        109.089        98.6843        34.9193
 15.6          284.364        162.72         158.997        37.3534
```

Program notes

(1) Example 5.1 is a modification of Example 3.3 to allow calculation of horizontal as well as vertical stresses. For each layer a ϕ' value as well as the level and unit weight are given. A variable AP is set to 0 for active stresses, 1 for passive.
(2) The vertical effective stress is calculated exactly as in 3.3, and multiplied by K_a or K_p to give σ_h'. The total horizontal stress σ_h is then calculated by adding the pore pressure u. The total and effective stresses are output in a table for different depths.

Examples 5.2 and 5.3 Coulomb wedge analysis

Write a program to calculate the active force on a retaining wall using Coulomb's method.

Coulomb's wedge analysis of a retaining wall is the oldest calculation in soil mechanics but is still a useful way of estimating the forces on a retaining wall. The analysis may take several different forms according to the soil and wall type; two examples only are given here. The first is for a wall supporting dry cohesionless material,

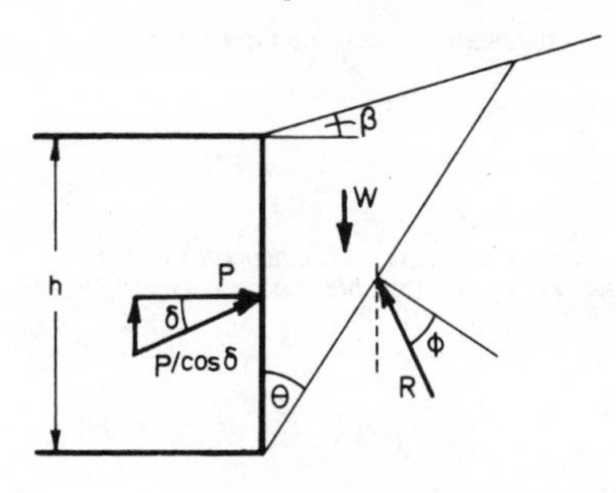

Figure 5.6 Wedge analysis in cohesionless soil

with the backfill sloping up from the top of the wall as shown in Figure 5.6. The second is for a wall supporting cohesive soil, including the effects of a tension crack and the possibility of the soil strength increasing with depth (see Figure 5.7).

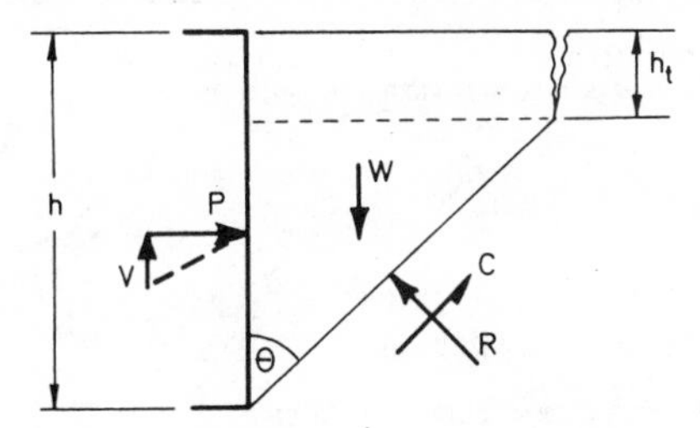

Figure 5.7 Wedge analysis in cohesive soil

```
LIST

10      REM EXAMPLE 5.2. COULOMB WEDGE ANALYSIS OF ACTIVE RETAINING WALL
20      REM FRICTIONAL SOIL
30      REM
40      PRINT "ENTER ANGLE OF INTERNAL FRICTION (PHI, DEGREES):"
50      INPUT PH
60      PH = PH * 3.14159 / 180.0
70      PRINT "ENTER UNIT WEIGHT OF SOIL (GAMMA, KN/M3):"
80      INPUT GA
90      PRINT "ENTER ANGLE OF WALL FRICTION (DELTA, DEGREES):"
100     INPUT DE
110     DE = DE * 3.14159 / 180.0
120     PRINT "ENTER HEIGHT OF WALL (H, M):"
130     INPUT H
140     PRINT "ENTER ANGLE OF BACKFILL UPWARDS FROM HORIZONTAL (BETA, DEG.):"
150     INPUT BE
160     BE = BE * 3.14159 / 180.0
170     REM
180     REM MAKE INITIAL GUESS FOR ANGLE OF FAILURE SURFACE THETA
190     REM
200     T = 3.14159 * 0.25 - PH * 0.5
210     PL = 0.0
220     TI = 0.1
230     REM
240     REM ENTER ITERATIVE CALCULATION
250     REM
260     W = (GA*H*H*0.5) * TAN(T) / (1.0-TAN(BE)*TAN(T))
270     P = W*TAN(0.5*3.14159-T-PH) / (1.0+TAN(DE)*TAN(0.5*3.14159-T-PH))
280     REM
```

```
290      REM CHECK ON IMPROVEMENT IN CALCULATION OF P
300      REM
310      IF P<PL THEN TI = - 0.5 * TI
320      IF ABS(TI)<0.001 GOTO 360
330      PL = P
340      T = T + TI
350      GOTO 260
360      PRINT "FORCE ON WALL PER UNIT LENGTH (KN/M): ",P
370      PRINT "CRITICAL ANGLE OF FAILURE SURFACE (DEG): ",180.0*T/3.14159
380      STOP
390      END

READY

RUN

ENTER ANGLE OF INTERNAL FRICTION (PHI, DEGREES):
? 30
ENTER UNIT WEIGHT OF SOIL (GAMMA, KN/M3):
? 20
ENTER ANGLE OF WALL FRICTION (DELTA, DEGREES):
? 0
ENTER HEIGHT OF WALL (H, M):
? 10
ENTER ANGLE OF BACKFILL UPWARDS FROM HORIZONTAL (BETA, DEG.):
? 0
FORCE ON WALL PER UNIT LENGTH (KN/M):        333.331
CRITICAL ANGLE OF FAILURE SURFACE (DEG):     30.0895

READY

RUN

ENTER ANGLE OF INTERNAL FRICTION (PHI, DEGREES):
? 38.5
ENTER UNIT WEIGHT OF SOIL (GAMMA, KN/M3):
? 19.7
ENTER ANGLE OF WALL FRICTION (DELTA, DEGREES):
? 11.0
ENTER HEIGHT OF WALL (H, M):
? 10.0
ENTER ANGLE OF BACKFILL UPWARDS FROM HORIZONTAL (BETA, DEG.):
? 7.2
FORCE ON WALL PER UNIT LENGTH (KN/M):        225.153
CRITICAL ANGLE OF FAILURE SURFACE (DEG):     28.5253
```

Program notes

(1) The program in Example 5.2 carries out an analysis of an active retaining wall supporting a dry frictional material ($c' = 0$). The wall is vertical and the angle of the backfill at β to the horizontal (see Figure 5.3). The angle of friction of the soil is ϕ' and it is assumed that the friction angle on the wall is δ ($\delta \leqslant \phi'$). A trial failure surface at angle θ is assumed and the value of P for equilibrium of the wedge may be calculated as:

$$P = \frac{W \tan\left(\frac{\pi}{2}-\theta-\phi\right)}{1 + \tan\delta \tan\left(\frac{\pi}{2}-\theta-\phi\right)} \tag{5.8}$$

where

$$W = \frac{\gamma h^2}{2} \frac{\tan\theta}{(1-\tan\beta\tan\theta)} \tag{5.9}$$

The value of θ is then varied until the most critical surface (i.e. the one which maximises P) is found.

(2) In Example 5.2 P is maximized in the following way. The program starts by inputting the data in the usual way and then by setting θ to $(\pi/4 - \phi'/2)$, i.e. the value it would take if $\delta = 0$ and $\beta = 0$.

Two additional values are set in lines 210 and 220; PL is initialized as zero, and TI (which will be used as an increment in θ) as 0.1 radians. A value of P is calculated for $\theta = (\pi/4 - \phi'/2)$ and at line 310 it is compared with PL. Since P is positive and PL = 0, TI remains unchanged, PL is reset to the value of P, T incremented by TI and the program returns to line 260 to recalculate P for a new θ value.

This time when line 310 is encountered the new value of P may be larger or smaller than the previous value. If it is larger the changes in P are moving in the right direction, so TI remains unchanged, if it is smaller then the step in θ is halved and reversed by TI = −0.5 * TI and so θ will be changed back towards the better values. The halving of the step size means that the value of θ gradually homes in to give the maximum P value (see Figure 5.8). At some stage this procedure must be stopped, and this is done by line 320, which detects when the increment in θ has become sufficiently small so that no further optimization is necessary, and forces the program to print the pressure and the critical angle (note the conversion from radians to degrees). This form of search for an optimum value is a convenient method which can be used in many different applications.

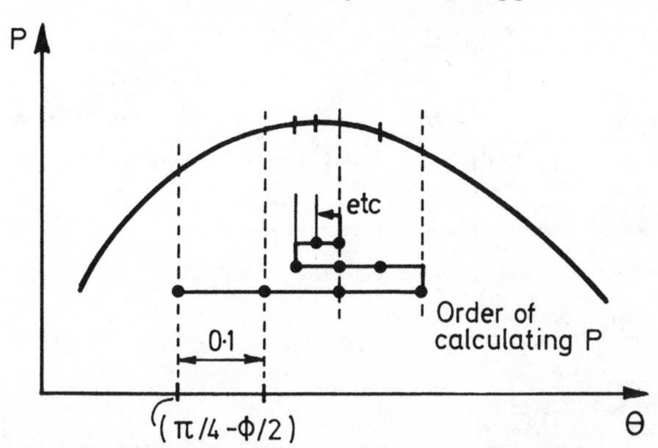

Figure 5.8 Routine for maximization of active force

(3) Two cases of running of Example 5.2 are given. The first uses $\phi' = 30°$ and $\delta = \beta = 0$, and for this case it is easy to show analytically that the critical value of θ is 30° and that $P = \gamma h^2/6$. The results of the program confirm that it is running correctly. This is an important stage in testing any program — several runs should be made using data for which the answer can be checked easily by hand.

(4) The second run shows a more realistic set of data, for which the hand calculation would be very tedious.

```
LIST

10       REM EXAMPLE 5.3. COULOMB WEDGE ANALYSIS OF ACTIVE RETAINING WALL
20       REM COHESIVE SOIL WITH INCREASE OF CU WITH DEPTH, TENSION CRACK
30       REM
40       PRINT "ENTER COHESION AT SOIL SURFACE (CO, KN/M2):"
50       INPUT CO
60       PRINT "ENTER INCREASE OF COHESION WITH DEPTH (C1, KN/M3):"
70       INPUT C1
80       PRINT "ENTER UNIT WEIGHT OF SOIL (GAMMA, KN/M3):"
90       INPUT GA
100      PRINT "ENTER WALL ADHESION AT TOP (AO, KN/M2):"
110      INPUT AO
120      PRINT "ENTER INCREASE OF WALL ADHESION WITH DEPTH (A1, KN/M3):"
130      INPUT A1
140      PRINT "ENTER HEIGHT OF WALL (H, M):"
150      INPUT H
160      PRINT "IS TENSION CRACK TO BE INCLUDED (Y/N)"
170      INPUT T$
180      IF T$="N" GOTO 240
190      IF T$<>"Y" GOTO 160
200      PRINT "MAY TENSION CRACK BE WATER FILLED (Y/N)"
210      INPUT W$
220      IF W$="N" GOTO 240
230      IF W$<>"Y" GOTO 200
240      REM
250      REM MAKE INITIAL GUESS FOR ANGLE OF FAILURE SURFACE THETA
260      REM
270      T = 3.14159 * 0.25
280      PL = 0.0
290      TI = 0.1
300      REM
310      REM ENTER ITERATIVE CALCULATION
320      REM
330      IF T$="N" GOTO 360
340      TC = 2*CO / (GA - 2*C1)
350      GOTO 370
360      TC = 0
370      W = GA * (0.5 * ((H - TC)^2) * TAN(T) + (H - TC) * TAN (T) * TC)
380      VW = (H - TC) * (AO + 0.5 * A1 * (H + TC))
390      C = (H - TC) * (CO + 0.5 * C1 * (H + TC)) / COS(T)
400      P = ((W - VW) / TAN(T)) - C / SIN(T)
410      IF T$="N" GOTO 430
420      IF W$="Y" THEN P = P + 9.81 * TC * TC * 0.5
430      REM
440      REM CHECK ON IMPROVEMENT IN CALCULATION OF P
450      REM
460      IF P<PL THEN TI = - 0.5 * TI
470      IF ABS(TI)<0.001 GOTO 510
480      PL = P
490      T = T + TI
500      GOTO 330
510      PRINT "FORCE PER UNIT LENGTH ON WALL (KN/M): ",P
520      PRINT "CRITICAL ANGLE OF FAILURE SURFACE (DEG): ",180.0*T/3.14159
530      IF T$="Y" THEN PRINT "DEPTH OF TENSION CRACK (M): ",TC
540      PRINT " "
550      PRINT "ENTER 0 TO STOP, 1 FOR NEW DATA:"
560      INPUT I
570      PRINT " "
580      IF I=1 GOTO 40
590      STOP
600      END

READY

RUN
```

```
ENTER COHESION AT SOIL SURFACE (CO, KN/M2):
? 20.5
ENTER INCREASE OF COHESION WITH DEPTH (C1, KN/M3):
? 3.1
ENTER UNIT WEIGHT OF SOIL (GAMMA, KN/M3):
? 19.7
ENTER WALL ADHESION AT TOP (AO, KN/M2):
? 7.0
ENTER INCREASE OF WALL ADHESION WITH DEPTH (A1, KN/M3):
? 1.0
ENTER HEIGHT OF WALL (H, M):
? 14.5
IS TENSION CRACK TO BE INCLUDED (Y/N)
? Y
MAY TENSION CRACK BE WATER FILLED (Y/N)
? N
FORCE PER UNIT LENGTH ON WALL (KN/M):        719.069
CRITICAL ANGLE OF FAILURE SURFACE (DEG):     49.2077
DEPTH OF TENSION CRACK (M):                  3.03704

ENTER 0 TO STOP, 1 FOR NEW DATA:
? 1

ENTER COHESION AT SOIL SURFACE (CO, KN/M2):
? 20.5
ENTER INCREASE OF COHESION WITH DEPTH (C1, KN/M3):
? 3.1
ENTER UNIT WEIGHT OF SOIL (GAMMA, KN/M3):
? 19.7
ENTER WALL ADHESION AT TOP (AO, KN/M2):
? 0
ENTER INCREASE OF WALL ADHESION WITH DEPTH (A1, KN/M3):
? 0
ENTER HEIGHT OF WALL (H, M):
? 14.5
IS TENSION CRACK TO BE INCLUDED (Y/N)
? Y
MAY TENSION CRACK BE WATER FILLED (Y/N)
? N
FORCE PER UNIT LENGTH ON WALL (KN/M):        886.942
CRITICAL ANGLE OF FAILURE SURFACE (DEG):     45.0895
DEPTH OF TENSION CRACK (M):                  3.03704

ENTER 0 TO STOP, 1 FOR NEW DATA:
? 1

ENTER COHESION AT SOIL SURFACE (CO, KN/M2):
? 20.5
ENTER INCREASE OF COHESION WITH DEPTH (C1, KN/M3):
? 3.1
ENTER UNIT WEIGHT OF SOIL (GAMMA, KN/M3):
? 19.7
ENTER WALL ADHESION AT TOP (AO, KN/M2):
? 7.0
ENTER INCREASE OF WALL ADHESION WITH DEPTH (A1, KN/M3):
? 1.0
ENTER HEIGHT OF WALL (H, M):
? 14.5
IS TENSION CRACK TO BE INCLUDED (Y/N)
? N
FORCE PER UNIT LENGTH ON WALL (KN/M):        632.818
CRITICAL ANGLE OF FAILURE SURFACE (DEG):     49.2077

ENTER 0 TO STOP, 1 FOR NEW DATA:
? 0
```

Program notes

(1) Example 5.3 is a program for Coulomb wedge analysis of an active wall in a cohesive soil. In its simplest form this calculation is rather trivial, but the program allows two extra sophistications to be made. Firstly it allows a tension crack (see Figure 5.3) which may exist in the ground up to a depth of $2c_u/\gamma$. The presence of a tension crack often results in a more critical failure mechanism, especially when the crack is filled with water (e.g. after a period of heavy rain). The second complication is that the strength of the soil is allowed to be a linear function of depth — a situation which is frequently encountered in practice. Similarly the adhesion on the wall is a linear function of depth.
(2) The structure of the program is exactly as in Example 5.2, and only the details of the calculation of P need be given here.

If the strength at the ground surface is c_0 and the rate of increase of strength c_1 then the tension crack depth is calculated from

$$h_t = \frac{2(c_0 + h_t c_1)}{\gamma} \tag{5.10}$$

The variable h_t is set therefore to $\frac{2c_0}{\gamma - 2c_1}$ if a tension crack is allowed and zero if it is not.

For a failure surface at angle θ the weight of the moving block is easily calculated as:

$$W = \frac{\gamma}{2}(h-h_t)^2 \tan\theta + \gamma(h-h_t)h_t \tan\theta \tag{5.11}$$

The shear force on the failure surface must be calculated using the average strength along the surface, thus

$$C = \frac{(h-h_t)[c_0 + \frac{1}{2}(h_t + h)c_1]}{\cos\theta} \tag{5.12}$$

and similarly the shear force on the wall as:

$$V = (h-h_t)[a_0 + \tfrac{1}{2}(h_t + h)a_1] \tag{5.13}$$

Finally P is calculated by resolving forces parallel to the failure surface:

$$P = \frac{W-V}{\tan\theta} - \frac{C}{\sin\theta} \tag{5.14}$$

If the tension cracks are water filled then the water pressure is added to P.

(3) The example runs for 5.3 show the analysis of a single wall for several different cases. The first is the analysis of a wall using dry tension cracks. In the second the effect of wall adhesion is eliminated and the pressure increases. In the third case the tension crack is excluded and a lower pressure is obtained.

Example 5.4 Stability of propped cantilever retaining wall

Write a program to analyse the propped cantilever retaining wall shown in Figure 5.9 so as to determine the value of the prop force and the overall factor of safety of the wall against failure by rotation of the wall about the prop. The soil is dry and cohesionless, the wall is of height H and the sheet piles of which it is constructed are driven to a depth D below the base of the wall.

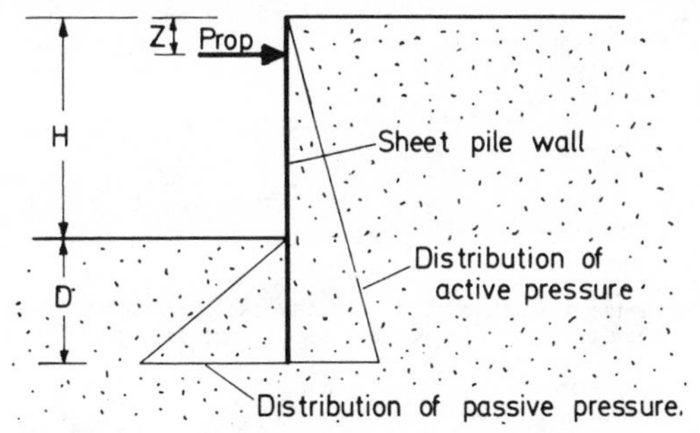

Figure 5.9 Propped cantilever retaining wall

If the fully active Rankine pressure distribution is assumed on the active side the force per unit length of wall will be $\frac{1}{2}\gamma(H+D)^2K_a$, and similarly the force on the passive side will be $\frac{1}{2}\gamma D^2K_p$. The difference between the two gives the prop force (see problem 5.4). Taking moments about the position of the prop, the factor of safety may be defined as the ratio of the moment of the passive resisting force to the moment of the active force. Note that because of the triangular pressure distribution in Rankine's calculation the active force acts at a distance $\frac{2}{3}(D+H)$ from the top of the wall and the passive force at a distance $(H+\frac{2}{3}D)$.

The program allows for the prop to be positioned below the top of the wall, but note that the distance Z should be only a small fraction of H or the assumption of the linear active pressure distribution may be inaccurate. Note also that the discussion of wall movements in section 5.5 is relevant to this problem. A simple Rankine pressure distribution will be obtained only if the prop yields sufficiently to allow fully active conditions to develop. On the passive side the maximum pressures that could be developed would be much greater if friction on the wall were taken into account; however large movements would be necessary to develop such pressures and restricting the pressures to those calculated for $\delta = 0$ is a commonly used method of ensuring that the movements under working conditions will not be excessive.

```
LIST

10      REM EXAMPLE 5.4. OVERALL STABILITY OF A SMOOTH PROPPED CANTILEVER
20      REM RETAINING WALL IN DRY FRICTIONAL SOIL
30      REM
40      PRINT "ENTER HEIGHT OF RETAINING WALL (EXCLUDING EMBEDMENT) (M)"
50      INPUT H
60      PRINT "ENTER DEPTH OF EMBEDMENT (M)"
70      INPUT D
80      PRINT "ENTER DISTANCE BELOW TOP OF WALL OF PROP (M)"
90      INPUT ZP
100     PRINT "ENTER BULK UNIT WEIGHT OF SOIL (KN/M3)"
110     INPUT GA
120     PRINT "ENTER ANGLE OF FRICTION (DEGREES)"
130     INPUT PH
140     PH = PH * 3.14159 / 180.0
150     KA = (1 - SIN(PH)) / (1 + SIN(PH))
160     KP = 1 / KA
170     TH = H + D
180     PA = 0.5 * GA * TH * TH * KA
190     PP = 0.5 * GA * D * D * KP
200     P = PA - PP
210     MA = PA * (TH/1.5 - ZP)
220     MP = PP * (H + D/1.5 - ZP)
230     FS = MP / MA
240     PRINT "PROP FORCE REQUIRED PER UNIT LENGTH ",P," KN/M"
250     PRINT "FACTOR OF SAFETY ON MOMENTS ABOUT PROP IS",FS
260     PRINT "ANOTHER CALCULATION ";
270     INPUT C$
280     IF C$="Y" GOTO 40
290     STOP
300     END

READY

RUN

ENTER HEIGHT OF RETAINING WALL (EXCLUDING EMBEDMENT) (M)
? 11.2
ENTER DEPTH OF EMBEDMENT (M)
? 4.6
ENTER DISTANCE BELOW TOP OF WALL OF PROP (M)
? 1.2
ENTER BULK UNIT WEIGHT OF SOIL (KN/M3)
? 18.9
ENTER ANGLE OF FRICTION (DEGREES)
? 31.0
PROP FORCE REQUIRED PER UNIT LENGTH        130.458        KN/M
FACTOR OF SAFETY ON MOMENTS ABOUT PROP IS  1.15814
ANOTHER CALCULATION ? N
```

Program notes

(1) The program is a very simple one to illustrate the concept of a factor of safety in the analysis of a retaining wall. The general format of the program follows that of earlier examples.

(2) The Rankine coefficients are calculated in lines 150 and 160, the resulting forces and moments in lines 170 to 220 and the factor of safety in line 230.

PROBLEMS

(5.1) Making use of the fact that the integral between points 1 and 2 of a quantity y which varies from y_1 at x_1 to y_2 at x_2 is $\frac{1}{2}(y_1 + y_2)(x_2 - x_1)$, modify the program in Example 5.1 to calculate the total horizontal force, effective horizontal force and water pressure on an active or passive retaining wall of a given depth. If the ground water level is not at the level of a layer interface take care that the distribution of horizontal stress is integrated correctly.
(5.2) Add the appropriate lines to the programs in Examples 5.2 and 5.3 so that active or passive retaining walls may be selected for analysis. It will probably be most convenient to make use of a variable which is set to $+1$ for an active wall and -1 for a passive wall.
(5.3) In order to check the sensitivity of the value of the pressure on a retaining wall to the value of θ, add an extra facility to one of the programs in Examples 5.2 or 5.3 to provide the following information. As well as printing the optimum θ value and maximum P value, P should also be calculated at angles $(\theta_{opt} + \alpha)$, where α takes values $-10°$, $-5°$, $-2°$, $-1°$, $1°$, $2°$, $5°$ and $10°$. It may be convenient to move lines 260 and 270 of program 5.2 (330 to 420 of program 5.3) to a separate subroutine. This technique illustrates how a computer can be used by an engineer to gain extra insight into a problem by carrying out several rapid calculations which give some additional information about a problem — information which would probably not be obtained if the calculation had to be done by hand.
(5.4) The program in Example 5.4 is deliberately simple and could be improved in a number of ways. As a start, the calculated prop force is only correct when the passive pressure has reached its maximum value. Under working conditions the passive pressures will be smaller, just sufficient to give equilibrium of moments about the prop, and the prop force must be larger. Modify the program to calculate the prop force under working conditions.

The problem for the design engineer will usually be to determine the necessary embedment depth D, for a particular height of wall and known soil conditions, to give the required factor of safety. Change the input routines so that repeat calculations may first be made with different values of D but without having to input each time the unchanged values of H, Z, γ and ϕ. The facility to proceed to further calculations in which any or all of the variables may be changed should be retained.
(5.5) Walls of the type shown in Figure 5.9 are frequently used in harbour and river training schemes, when the ground conditions are commonly layered alluvial soils and the level of the water table on

either side of the wall may be different. For such conditions the active and passive pressures may be obtained from a calculation of the kind given in Example 5.1. Combine the programs of Example 5.1 and 5.4 to analyse such a situation.

Alternatively the wall may be used to retain the toe of a spoil heap or road embankment. Rewrite the program of Example 5.4 to allow for a sloping surface for the backfill behind the wall. Calculate the force on the wall using Coulomb's method, using ideas from Example 5.2, which will also allow an angle of wall friction δ to be included without difficulty. It is a reasonable approximation to assume that the resultant active force still acts at a depth of $\frac{2}{3}(H + D)$. Note that if wall friction is included there will also be a vertical resultant force on the wall which should also be calculated and output by the program.

(5.6) Figure 5.10 shows two common types of 'gravity' retaining walls which rely on their own mass for their stability. The first is a thick wall of mass concrete or masonry, the second an L-shaped wall of reinforced concrete which may be analysed as a block comprising the wall and the soil resting on the base of the wall. Movement of the wall is resisted by friction or cohesion along the base of the wall and by a distribution of pressure under the wall which is assumed to be trapezoidal in shape, as shown in Figure 5.10. Write a program to determine the active force on the back of the wall or block, the factor of safety against sliding, and the bearing pressures under the base of the wall.

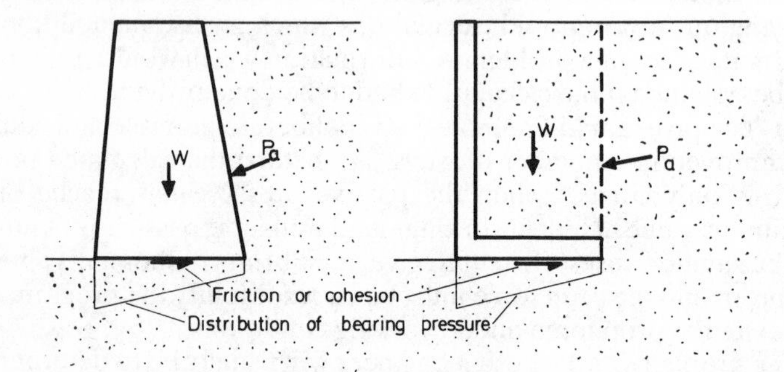

Figure 5.10 Gravity retaining walls

(5.7) Any of the programs developed above from Example 5.4 or Program (5.6) may be extended to calculate the forces (direct force, shear force and bending moment) within the structures of the wall. Make such an addition to one of the programs, with reference to Chapter 5 of *BASIC Stress Analysis* by M. J. Iremonger, a companion volume in this series.

Chapter 6

Slope stability

ESSENTIAL THEORY

6.1 Problems of slope stability

Many problems in soil mechanics involve the stability of slopes. These may be natural slopes, whose stability is perhaps threatened by natural erosion or man-made excavation; the side slopes of cuttings and excavations; or the sides of embankments and spoil heaps.

Methods of calculation and models of soil behaviour used are similar to those discussed in section 5.1. Again, two particular conditions may be identified; short term undrained conditions in fine grained soils for which total stress analyses using an undrained shear strength c_u for the soil ($\phi' = 0$) are appropriate; and long term conditions for which effective stress analyses are appropriate. In the latter case, especially for natural or cut slopes in heavily overconsolidated clay, a value for c' as well as ϕ' may be included on the basis of laboratory testing or back analysis of previous failures. However it will be found that the value selected for c' often has a crucial effect on the outcome of the calculations, and values should be chosen with care and without optimism.

Slope problems may involve mean stress either increasing or decreasing with time. Consideration must be given to the type of soil and the stress changes involved to determine whether short term or long term conditions will prove more critical. In many cases it may be wise to check both conditions, taking due account of changes in pore water pressure with changes in the ground water regime accompanying the work. During the construction of an embankment there may be a dynamic balance between the generation of pore pressure due to increasing total stresses and their dissipation by drainage, and account must then be taken of the expected pore pressures at different times and stages of construction to ensure that stability is maintained at all times.

6.2 Infinite slopes

A useful introduction to the stability of slopes may be made by considering slopes that are sufficiently long that they may be considered as infinite in extent. Any one section of the slope must

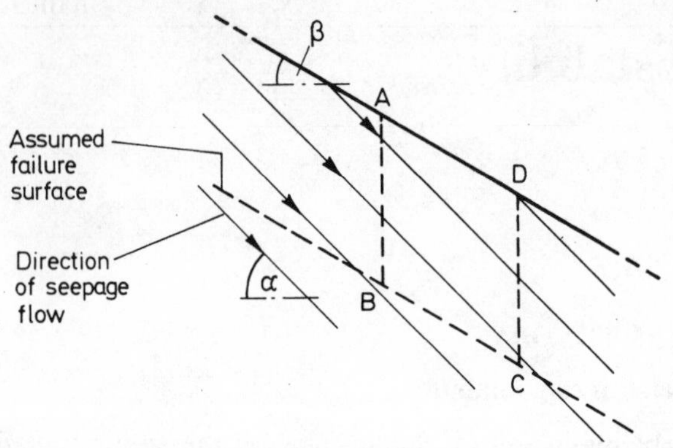

Figure 6.1 Fully saturated infinite slope, variable direction of seepage

then be in the same state as any other. Figure 6.1 shows a fairly general condition in which a slope at angle β of soil with $c' = 0$ is fully saturated, with steady seepage occurring in the direction defined by the angle α. Note that α may be positive or negative and covers the range of conditions from percolation into the slope to flow parallel to the slope to seepage out of the slope.

A limit equilibrium analysis of the typical block of soil ABCD gives an expression for the factor of safety F of the slope

$$F = \frac{\tan\phi'}{\tan\beta}\left[1 - \frac{\gamma_w(1 + \tan^2\beta)}{\gamma(1 + \tan\alpha\tan\beta)}\right] \tag{6.1}$$

where F is as before defined as the ratio of available to mobilised soil strength on the assumed failure plane, γ is the saturated unit weight of the soil and γ_w the unit weight of water.

A number of special cases may easily be deduced from Equation (6.1). For instance if the slope is dry rather than saturated the second term within the square brackets disappears, giving

$$F = \frac{\tan\phi'}{\tan\beta} \tag{6.2}$$

Such a slope is thus just stable ($F = 1$) when the slope angle is equal to the angle of internal friction ϕ'. This slope angle is known as the 'angle of repose' for dry granular material.

For vertical percolation of water into the slope $\alpha = 90°$ and Equation (6.1) again reduces to Equation (6.2). For flow parallel to the slope $\alpha = \beta$, and

$$F = \frac{\tan\phi'}{\tan\beta}\left[1 - \frac{\gamma_w}{\gamma}\right] \approx \frac{\tan\phi'}{2\tan\beta}, \text{ since } \gamma \approx 2\gamma_w \tag{6.3}$$

Note that in Equations (6.1) to (6.3) the factor of safety is independent of the depth assumed for the failure surface. Failure will tend to occur either on a plane of weaker soil within the mass or close to the surface where the changes needed to trigger a failure are most likely to occur.

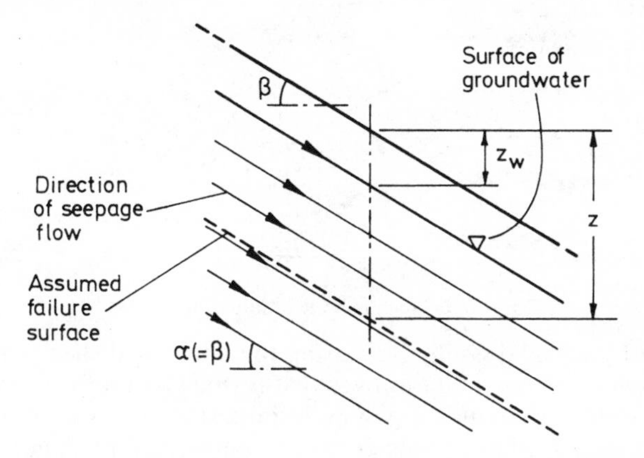

Figure 6.2 Infinite slope, depressed water table, seepage flow parallel to slope

If, however, the case is considered in which seepage is parallel to the slope but the water table is below the surface (Figure 6.2), the factor of safety becomes

$$F = \frac{\tan \phi'}{\tan \beta}\left[1 - \frac{\gamma_w}{\gamma} + \frac{\gamma_w z_w}{\gamma\ z}\right] \tag{6.4}$$

where z_w is the depth to the water table and z the depth to the assumed failure plane. F clearly decreases as z increases, and failure will try to occur as deep as possible, perhaps at the interface between the soil and underlying rock. Such conditions are a reasonable approximation to those often occurring in natural slopes in upland areas of high rainfall.

A similar analysis may be performed for the short term stability of an infinite slope in cohesive soil, giving

$$F = \frac{c_u}{\gamma z \cos \beta \sin \beta} = \frac{2c_u}{\gamma z \sin 2\beta} \tag{6.5}$$

In practice $c_u/\gamma z$ tends to decrease with z and failures again will try to occur as deep as possible.

When failures occur at depth, as in these last two cases, the assumption of an infinite slope of homogeneous material is unlikely to be a reasonable approximation to the real situation. Many real slips occur along roughly circular failure surfaces and this assumption is made in the standard methods of analysis covered in the following sections.

6.3 Total stress analyses

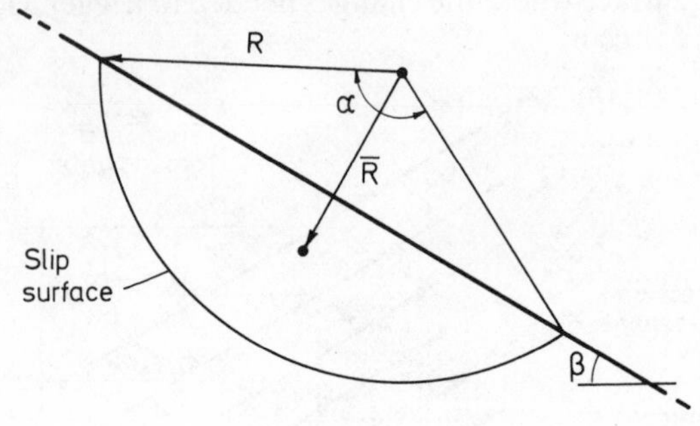

Figure 6.3 Circular slip in infinite slope

As usual a total stress analysis using the undrained shear strength of the soil is appropriate for investigating the short term stability of cohesive soil. A circular slip in an infinite slope is shown in Figure 6.3. An 'upper bound' analysis may be performed by equating the work done by gravity to the work done on the slip surface for a small rotation of the body of soil about the centre of rotation. This gives a factor of safety for the slope

$$F = \frac{2c_u\alpha}{\gamma\bar{R}(\alpha - \sin\alpha)\sin\beta} \qquad (6.6)$$

where $\bar{R}$ is the distance of the centre of mass of the sliding soil from the centre of rotation. The distance of the centre of rotation from the slope (and hence α and $\bar{R}$) may be varied to determine the minimum value of F.

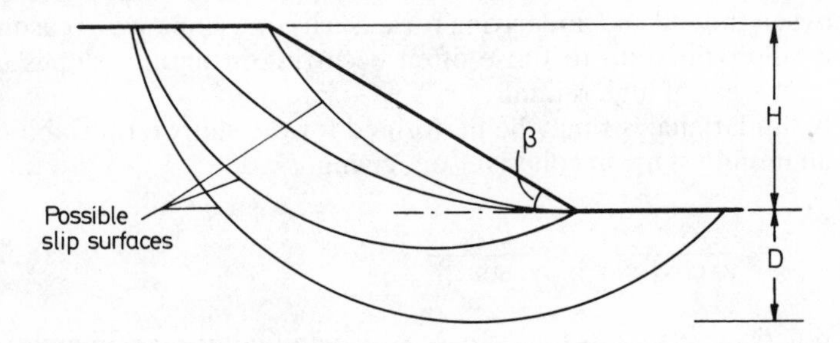

Figure 6.4 Slip surfaces for finite slopes

In practice the slope will not normally be infinite and failure may occur more easily by one of the mechanisms shown in Figure 6.4, depending on the geometry of the slope and the presence or not of a much stronger stratum below the slope. The stability of the slope may be further reduced by the presence of a tension crack at the top.

A similar type of analysis may be used (a limit-equilibrium analysis gives identical results), but a number of different centres and radii to the circles must be tried to determine the most critical surface and corresponding lowest factor of safety. A computer analysis is clearly ideal for such a repetitive calculation. Layers of soil with different densities and strengths may be included if appropriate, though the possibility of non-circular slips (see section 6.6) may then have to be considered; more complex slope geometries may also be analysed. However for such problems the method of slices discussed in the next section is normally used.

By invoking ideas of geometrical similarity Taylor expressed the stability of slopes in homogeneous soil in terms of a stability number N_s where

$$N_s = \frac{C_u}{F\gamma H} \qquad (6.7)$$

Taylor produced charts relating N_s to ϕ_u and slope angle β for various values of D (see Figure 6.4). Normally $\phi_u = 0$, and if D is large and β less than 53° then N_s has a constant value of 0.18. For other conditions reference should be made to the charts, which are reproduced in many textbooks. N_s increases for $\beta > 53°$, and is reduced for $\beta < 53°$ when the depth of slip is limited by a firm stratum.

6.4 Analysis by the methods of slices

The most general methods for analysing slopes are the methods of slices due to Fellenius (the Swedish method) and Bishop. Both may be used either for total stress analysis using undrained parameters ϕ_u, c_u or effective stress analysis using drained parameters ϕ', c' plus a knowledge of the pore pressures present in the soil.

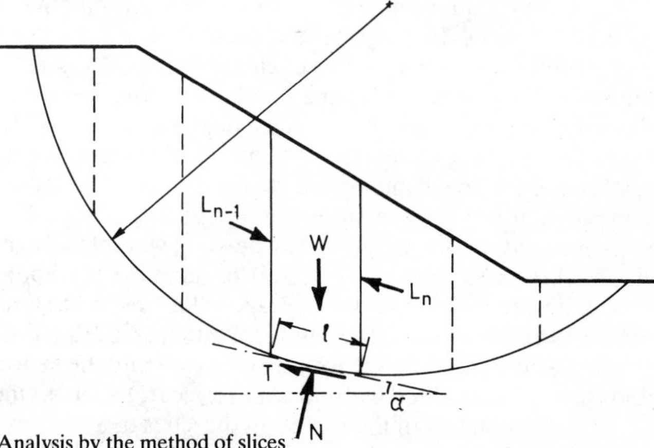

Figure 6.5 Analysis by the method of slices

A slip surface is assumed; this may be any shape but is commonly taken to be circular from the evidence from real slip failures. The sliding mass of soil is divided by vertical planes into a number of slices, as shown in Figure 6.5, and the equilibrium of these slices considered. The forces on either side of a slice, L_n and L_{n-1} in Figure 6.5, are approximately equal and in the Swedish method are assumed to cancel so that the total normal force on the base of each slice (N) may be determined explicitly as equal to $W \cos \alpha$. The factor of safety is defined as the ratio between the sum of all the resisting forces on the base of the slices and the sum of the components of the gravity forces parallel to the base of the slices, so that

$$F = \frac{1}{\Sigma W \sin\alpha} \Sigma [cl + (W\cos\alpha - ul) \tan \phi] \qquad (6.8)$$

u is the mean pore pressure on the base of a slice and this term is obviously only included in an effective stress analysis, in which case c and ϕ also have their effective (drained) values. It is sometimes convenient to include the pore pressure in terms of a pore pressure ratio r_u defined for each slice by

$$r_u = \frac{u}{\gamma z} \qquad (6.9)$$

u being the mean pore pressure and γz the mean overburden pressure at the base of the slice.

Equation (6.8) may then be rewritten as

$$F = \frac{1}{\Sigma W \sin\alpha} \Sigma [cl + W(\cos \alpha - r_u \sec \alpha) \tan \phi] \qquad (6.10)$$

For either Equation (6.8) or (6.10) the summations are made over all the slices. In general the value of r_u is different for each slice and its determination requires evaluation of u. For long term problems involving steady seepage conditions pore pressures may be determined from a flow net — see Chapter 6 of *BASIC Hydraulics* by P. D. Smith, a companion volume in this series, or a standard textbook on seepage and groundwater flow. For methods of determining r_u for pore pressures set up during construction of an embankment, or due to rapid drawdown of water level against a soil slope, reference should be made to standard textbooks on soil mechanics. In many cases the average value of r_u for a slope is determined and used as a constant value for all slices in Equation (6.10).

The Swedish method is known to overestimate the factor of safety and Bishop's simplified method is more accurate. In this method the vertical components of the forces on the sides of a slice are assumed to cancel, but not the horizontal components. Instead the assumption is made that each slice has the same factor of safety. The factor of safety is then given by

$$F = \frac{1}{\Sigma W \sin\alpha} \Sigma \left[\{cb + (W - ub)\tan\phi\} \left\{ \frac{\sec\alpha}{1 + \frac{\tan\alpha \tan\phi}{F}} \right\} \right] \quad (6.11)$$

or

$$F = \frac{1}{\Sigma W \sin\alpha} \Sigma \left[\{cb + W(1 - r_u)\tan\phi\} \left\{ \frac{\sec\alpha}{1 + \frac{\tan\alpha \tan\phi}{F}} \right\} \right] \quad (6.12)$$

where $b = l\cos\alpha$ and the same comments apply as for the Swedish method.

In Equations (6.11) and (6.12) the factor of safety appears within as well as outside the summation and cannot therefore be determined directly. Instead an iterative procedure is used. An initial factor of safety (say 1.0) is guessed for insertion in the right-hand side of the equation and a value of F calculated. This value is then input on the right-hand side of the equation and a new value of F calculated. The procedure is repeated until a constant value of F is approached. The iterative process is in fact very efficient and after three or four iterations the factor of safety usually remains the same to the third decimal place.

6.5 Stability coefficients

For a slope of homogeneous soil for which the possibility of tension cracks is ignored and r_u is taken to be a constant the factor of safety is linearly related to r_u, thus

$$F = m - nr_u \quad (6.13)$$

where m and n are stability coefficients depending on $c'/\gamma H$, $\cot\beta$, ϕ' and D.

The stability coefficients m and n may be obtained from charts published by Bishop and Morgenstern and reproduced in Reference 6. This procedure allows an approximate analysis using mean values for parameters to be made very quickly; by varying the values of the parameters between likely limits the sensitivity of the value of F to the different factors may be easily investigated.

6.6 Non-circular slips

In some cases neither the 'slab' slides predicted by the analyses of section 6.2 nor the circular slips assumed in section 6.3 to 6.5 are a reasonable model of the real situation. Typical examples might be sliding of a natural soil slope on an underlying rock surface, or lateral spreading of an embankment underlain by a layer of very weak clay, as shown in Figure 6.6. Such slips may often be analysed by approximating the slip surface to a series of planes, as indicated.

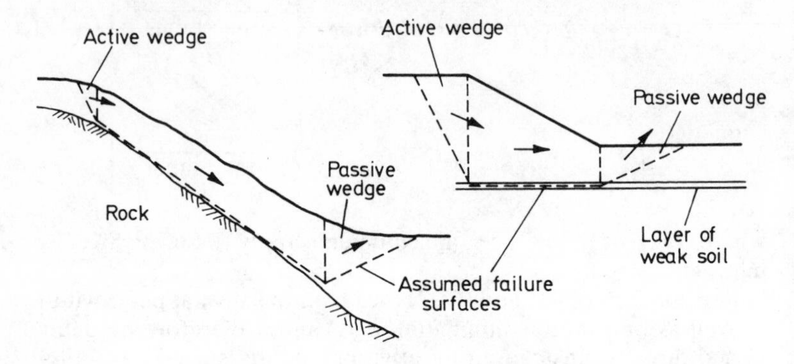

Figure 6.6 Wedge analysis of non-circular slips

Effectively the central block of soil slides under the action of its own weight and the active pressure from the upper 'Coulomb wedge', sliding being resisted by shear forces along its base plane and the passive pressure from the lower Coulomb wedge.

Such motion generally requires relative vertical motion to take place between the various blocks of soil, so that the active and passive forces are those that would be determined for $\delta = \phi'$ or $c_a = c_u$ by the methods outlined in section 5.4. Using the Rankine active and passive pressures, for which $\delta = c_a = 0$, will always give an underestimate (safe) of the factor of safety for the slope.

6.7 Monitoring of slopes

Geotechnical engineering can never be an exact science because of the variability and complexity of the material involved. Monitoring of construction is therefore frequently undertaken, to confirm the design methods used (and perhaps in time improve them) and ensure that the structure is performing satisfactorily. This is particularly true of earth slopes. Natural slopes sometimes have a very low factor of safety which may be further reduced by excavation or other work for new construction, while the stability of cut slopes may decrease with time as discussed above. Conversely embankments and earth dams, particularly when built on relatively soft foundation soils, usually become more stable as consolidation occurs. For economical construction they may therefore be built with initially very low factors of safety, which are acceptable in the short term provided a watchful eye is kept on the behaviour of the embankment and construction halted and remedial action taken if necessary.

As discussed earlier in this chapter the stability of a slope is sensitive to pore pressures in the soil, and the commonest form of instrumentation is the piezometer, which is used to measure pore pressure. In its simplest form, satisfactory in reasonably coarse and permeable soils, it consists of an open tube either installed in a borehole or built into an embankment during construction. Water rises within the tube until it is in equilibrium with the pore pressure at the base of the tube, and the level of the water may be detected by an electrical probe. In less permeable soils it would take too long for equilibrium to be established as water must flow out of or into the soil, and there is also a tendency for soil particles to be washed into the tube. In this case a piezometer tip, which consists of a porous element of relatively large surface area, is attached to a tube of small diameter. This may again be used as an open riser tube, or may be used in a closed system in which the tube is kept full of water and led to a pressure gauge below the level of the tip, so that the pore pressure may be determined from the gauge reading. When the soil is not saturated, but contains air or gas, care has to be taken to ensure that the pore-water, not pore-air, pressure is being measured and that air does not get into the measuring system. A double tube arrangement allows the tubes and tip to be flushed through with de-aired water to remove any air bubbles which may have collected in the system.

Many other types of piezometer have been developed, using transducers of some kind so that the pore pressure is measured in terms of pneumatic pressure, hydraulic pressure or an electrical signal. Details of these instruments, and of others mentioned below, may be found in *Foundation Instrumentation* by T. H. Hanna, Trans Tech Publications (1973).

In many cases the movements of soil slopes may be important, either as an indication of impending instability or as a measure of the progress of consolidation, and several types of instrument have been devised to measure movements in the soil. Surface movements may often be determined by accurate surveying techniques. Since slope movements are often predominantly horizontal in direction, perhaps the most useful instrument is the inclinometer, which provides a profile of horizontal movements with depth at a particular location. An inclinometer consists of a flexible tube which is installed approximately vertically in a borehole or built into an embankment. A probe can be lowered down the tube, running in guide channels, and measures changes in inclination of the tube. A typical instrument measures the inclination over a length of 0.5 m so that by taking readings every 0.5 m up the tube the full profile of the tube may be calculated (see worked Example 6.4). By comparing the profiles of different dates any movement of the embankment can be detected. A general deformation of the fill or subsoil may appear as a gradual bending or tilting of the tube, while the formation of a plane of sliding will cause an intense local deformation of the tube (see Figure 6.7).

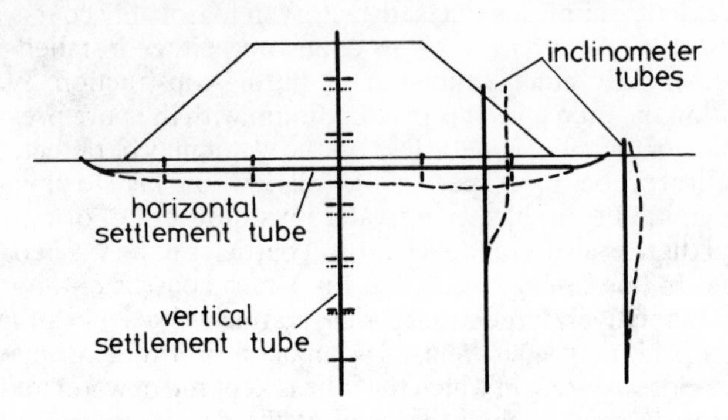

Figure 6.7 Typical instrumentation of embankment on soft ground

A rather similar idea is used to determine a settlement profile beneath an embankment. A flexible tube is laid in the ground before construction begins, as shown in Figure 6.7, but in this case the probe which may be drawn through the tube is sensitive to level changes and determines the settlement of the tube at measured distances along it. Horizontal soil movements along the tube may also be measured by fitting metal plates loosely around the tube at intervals, so that they slide along the tube as the soil moves. The probe is able to determine the position of the plates by the changes in electrical inductance as it passes a plate. A similar principle is applied to measurement of vertical movements by a vertical tube settlement gauge.

WORKED EXAMPLES

Examples 6.1, 6.2 and 6.3 Analysis of slopes by method of slices

The analysis of slopes is unusual in soil mechanics in that in practice it is carried out almost exclusively using a single method, the method of slices. Write programs based on this method, starting from the simplest possible program and then developing it to increasing levels of sophistication and usefulness.

```
LIST

10      REM EXAMPLE 6.1. SLOPE STABILITY ANALYSIS, SWEDISH METHOD OF SLICES
20      REM
30      DIM A(20),C(20),L(20),P(20),U(20),W(20)
40      REM
50      REM INPUT THE DATA
60      REM
70      PRINT "ENTER NUMBER OF SLICES"
```

```
80      INPUT N
90      PRINT "ENTER     C, PHI, ALPHA, L,  W,      U FOR EACH SLICE"
100     PRINT "UNITS KN/M2, DEG,   DEG, M, KN, KN/M2"
110     FOR I=1 TO N
120     PRINT "SLICE NUMBER",I
130     INPUT C(I),P(I),A(I),L(I),W(I),U(I)
140     A(I) = A(I) * 3.14159 / 180.0
150     P(I) = P(I) * 3.14159 / 180.0
160     NEXT I
170     REM
180     REM CALCULATE FACTOR OF SAFETY
190     REM
200     T=0.0
210     B=0.0
220     FOR I=1 TO N
230     T=T + C(I)*L(I) + TAN(P(I))*(W(I)*COS(A(I)) - U(I)*L(I))
240     B=B + W(I)*SIN(A(I))
250     NEXT I
260     F=T/B
270     PRINT "FACTOR OF SAFETY (SWEDISH METHOD) IS",F
280     STOP
290     END

READY

RUN

ENTER NUMBER OF SLICES
? 5
ENTER     C, PHI, ALPHA, L,  W,      U FOR EACH SLICE
UNITS KN/M2, DEG,   DEG, M, KN, KN/M2
SLICE NUMBER   1
? 25, 30, 55, 28, 3456, 0
SLICE NUMBER   2
? 25, 30, 37, 25, 8856, 0
SLICE NUMBER   3
? 25, 30, 23, 21, 9504, 0
SLICE NUMBER   4
? 25, 30, 10, 20.5, 7776, 0
SLICE NUMBER   5
? 25, 30, -6, 25.5, 4050, 0
FACTOR OF SAFETY (SWEDISH METHOD) IS       1.56436

READY

RUN

ENTER NUMBER OF SLICES
? 5
ENTER     C, PHI, ALPHA, L,  W,      U FOR EACH SLICE
UNITS KN/M2, DEG,   DEG, M, KN, KN/M2
SLICE NUMBER   1
? 25, 30, 55, 28, 3456, 108
SLICE NUMBER   2
? 25, 30, 37, 25, 8856, 221.4
SLICE NUMBER   3
? 25, 30, 23, 21, 9504, 237.6
SLICE NUMBER   4
? 25, 30, 10, 20.5, 7776, 194.4
SLICE NUMBER   5
? 25, 30, -6, 25.5, 4050, 81
FACTOR OF SAFETY (SWEDISH METHOD) IS       .680402
```

Program notes

(1) The program in Example 6.1 is a simple program to carry out an analysis using the Swedish method of slices. Up to 20 slices may be

used, the properties for each being input by the routine between lines 90 and 160 and stored as subscripted variables in arrays dimensioned in line 30.

(2) The properties c, ϕ, W, l, α and u have to be entered explicitly for each slice and have therefore to be determined separately by hand before using the program. Note that units must be consistent.

(3) The program simply calculates for each slice and sums in lines 230 and 240 the expressions in the numerator and denominator of the right-hand side of Equation (6.8) and determines the factor of safety in line 260 by dividing the former by the latter.

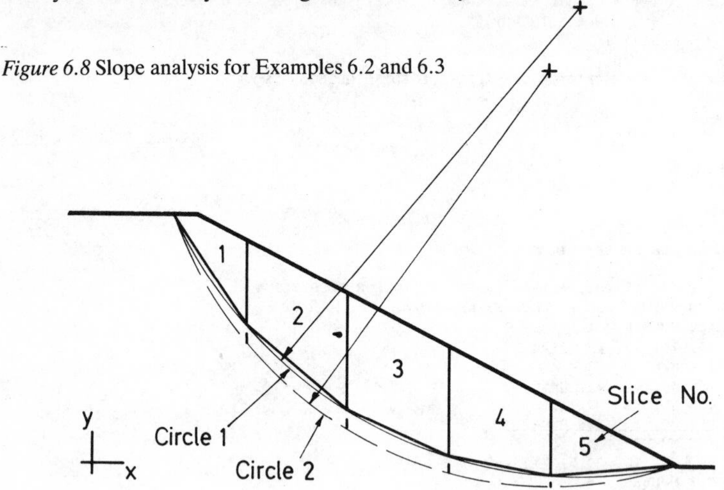

Figure 6.8 Slope analysis for Examples 6.2 and 6.3

(4) The example runs of the program are for the slope shown in Figure 6.7. The failure surface is assumed to be a circular arc, approximated by chords forming the bases of five slices in which the sliding mass of soil is divided. The analyses are in terms of effective stresses, with soil parameters $c' = 25$ kN/m^2 and $\phi' = 30°$ throughout. In the first case the slope is assumed to be dry and $u = 0$ throughout. In the second case the slope is assumed to be saturated, and the values of u correspond to a water surface coinciding with the profile of the slope. This might represent the most critical conditions following heavy rain or rapid drawdown of the adjoining water level after the slope had been completely inundated. The factor of safety is greatly reduced, to less than unity, predicting failure of the slope under these conditions.

```
LIST

10      REM EXAMPLE 6.2. SLOPE STABILITY ANALYSIS, SIMPLIFIED BISHOP METHOD
20      REM
30      DIM A(20),C(20),L(20),P(20),U(20),W(20)
40      REM
50      REM INPUT THE DATA
60      REM
70      PRINT "ENTER NUMBER OF SLICES"
80      INPUT N
```

```
90      PRINT "ENTER     C, PHI, ALPHA, L,  W,      U FOR EACH SLICE"
100     PRINT "UNITS KN/M2, DEG,   DEG, M, KN, KN/M2"
110     FOR I=1 TO N
120     PRINT "SLICE NUMBER",I
130     INPUT C(I),P(I),A(I),L(I),W(I),U(I)
140     A(I) = A(I) * 3.14159 / 180.0
150     P(I) = P(I) * 3.14159 / 180.0
160     NEXT I
170     REM
180     REM CALCULATE FACTOR OF SAFETY, INITIAL GUESS 1.0
190     REM
200     F=1.0
210     T=0.0
220     B=0.0
230     FOR I=1 TO N
240     G=1 + TAN(A(I))*TAN(P(I))/F
250     T=T + (C(I)*L(I) + TAN(P(I))*(W(I)/COS(A(I)) - U(I)*L(I)))/G
260     B=B + W(I)*SIN(A(I))
270     NEXT I
280     F=T/B
290     PRINT "FACTOR OF SAFETY (SIMPLIFIED BISHOP METHOD) IS",F
300     REM
310     REM OPTIONS ON RECALCULATING OR CHANGING DATA
320     REM
330     PRINT "ENTER -1    TO RECALCULATE"
340     PRINT "       0    TO EXIT"
350     PRINT "       1..N TO CHANGE DATA FOR A SLICE (GIVE SLICE NUMBER)"
360     INPUT J
370     IF J=-1 GOTO 210
380     IF J=0 THEN STOP
390     PRINT "ENTER NEW C, PHI, ALPHA, L,  W,     U FOR SLICE NUMBER",J
400     PRINT "UNITS KN/M2, DEG,   DEG, M, KN, KN/M2"
410     INPUT C(J),P(J),A(J),L(J),W(J),U(J)
420     A(J) = A(J) * 3.14159 / 180.0
430     P(J) = P(J) * 3.14159 / 180.0
440     GOTO 200
450     END

READY

RUN

ENTER NUMBER OF SLICES
? 5
ENTER     C, PHI, ALPHA, L,  W,     U FOR EACH SLICE
UNITS KN/M2, DEG,   DEG, M, KN, KN/M2
SLICE NUMBER   1
? 25, 30, 55, 28, 3456, 0
SLICE NUMBER   2
? 25, 30, 37, 25, 8856, 0
SLICE NUMBER   3
? 25, 30, 23, 21, 9504, 0
SLICE NUMBER   4
? 25, 30, 10, 20.5, 7776, 0
SLICE NUMBER   5
? 25, 30, -6, 25.5, 4050, 0
FACTOR OF SAFETY (SIMPLIFIED BISHOP METHOD) IS          1.57649
ENTER -1    TO RECALCULATE
       0    TO EXIT
       1..N TO CHANGE DATA FOR A SLICE (GIVE SLICE NUMBER)
? -1
FACTOR OF SAFETY (SIMPLIFIED BISHOP METHOD) IS          1.698
ENTER -1    TO RECALCULATE
       0    TO EXIT
       1..N TO CHANGE DATA FOR A SLICE (GIVE SLICE NUMBER)
? -1
FACTOR OF SAFETY (SIMPLIFIED BISHOP METHOD) IS          1.71536
```

```
ENTER -1     TO RECALCULATE
       0     TO EXIT
       1..N TO CHANGE DATA FOR A SLICE (GIVE SLICE NUMBER)
? -1
FACTOR OF SAFETY (SIMPLIFIED BISHOP METHOD) IS             1.71768
ENTER -1     TO RECALCULATE
       0     TO EXIT
       1..N TO CHANGE DATA FOR A SLICE (GIVE SLICE NUMBER)
FACTOR OF SAFETY (SIMPLIFIED BISHOP METHOD) IS             1.71799
? -1
ENTER -1     TO RECALCULATE
       0     TO EXIT
       1..N TO CHANGE DATA FOR A SLICE (GIVE SLICE NUMBER)
? 5
ENTER NEW C, PHI, ALPHA, L,   W,      U FOR SLICE NUMBER   5
UNITS KN/M2, DEG,    DEG, M, KN, KN/M2
? 0, 30, -6, 25.5, 4050, 0
FACTOR OF SAFETY (SIMPLIFIED BISHOP METHOD) IS             1.52347
ENTER -1     TO RECALCULATE
       0     TO EXIT
       1..N TO CHANGE DATA FOR A SLICE (GIVE SLICE NUMBER)
? -1
FACTOR OF SAFETY (SIMPLIFIED BISHOP METHOD) IS             1.6379
ENTER -1     TO RECALCULATE
       0     TO EXIT
       1..N TO CHANGE DATA FOR A SLICE (GIVE SLICE NUMBER)
? -1
FACTOR OF SAFETY (SIMPLIFIED BISHOP METHOD) IS             1.65531
ENTER -1     TO RECALCULATE
       0     TO EXIT
       1..N TO CHANGE DATA FOR A SLICE (GIVE SLICE NUMBER)
? -1
FACTOR OF SAFETY (SIMPLIFIED BISHOP METHOD) IS             1.65779
ENTER -1     TO RECALCULATE
       0     TO EXIT
       1..N TO CHANGE DATA FOR A SLICE (GIVE SLICE NUMBER)
? -1
FACTOR OF SAFETY (SIMPLIFIED BISHOP METHOD) IS             1.65814
ENTER -1     TO RECALCULATE
       0     TO EXIT
       1..N TO CHANGE DATA FOR A SLICE (GIVE SLICE NUMBER)
? 0
```

Program notes

(1) Bishop's method is used for the program in Example 6.2. The input is exactly the same as in Example 6.1, but an initial guess at the factor of safety of 1.0 is made in line 200. The factor of safety is then calculated in accordance with Equation (6.11) in lines 240 to 260.
(2) The option is then given, by entering −1 at line 360, of continuing with another iteration to calculate a more accurate value for the factor of safety by repeating lines 210 to 280. As the output from the program shows, five iterations are sufficient to give an approximately constant factor of safety. The data used are the same as for the first run in Example 6.1.
(3) An option is also given to allow the data for one slice to be changed, by entering the slice number at line 360. In the example given the effect of reducing the cohesion on the base of the last slice to zero is examined, and as expected the factor of safety is reduced.

```
LIST

10      REM EXAMPLE 6.3. SLOPE STABILITY ANALYSIS, SIMPLIFIED BISHOP METHOD
20      REM
30      DIM A(20),C(20),L(20),P(20),U(20),W(20)
40      DIM X(20),XB(20),YT(20),GA(20)
50      REM
60      REM INPUT THE DATA
70      REM
80      PRINT "ENTER NUMBER OF SLICES"
90      INPUT N
100     PRINT "ENTER X COORDINATE OF TOP OF SLOPE (M):"
110     INPUT X(0)
120     PRINT "ENTER BOTTOM AND TOP Y COORDS., LEFT SIDE OF TOP SLICE (M);"
130     INPUT YB(0),YT(0)
140     PRINT "ENTER WATER TABLE LEVEL AT TOP OF SLOPE (M):"
150     INPUT YW(0)
160     PRINT "ENTER DATA FOR EACH SLICE:"
170     GOSUB 1000
180     FOR I=1 TO N
190     GOSUB 2000
200     NEXT I
210     REM
220     REM CALCULATE DERIVED QUANTITIES
230     REM
240     FOR I=1 TO N
250     A(I) = ATN((YB(I)-YB(I-1)) / (X(I-1)-X(I)))
260     L(I) = SQR(((YB(I)-YB(I-1))^2) + ((X(I)-X(I-1))^2))
270     U(I) = 9.81 * 0.5 * (YW(I) + YW(I-1) - YB(I) -YB(I-1))
280     IF U(I)<0 THEN U(I)=0
290     W(I) = GA(I) * (X(I)-X(I-1)) * 0.5 * (YT(I)+YT(I-1)-YB(I)-YB(I-1))
300     NEXT I
310     REM
320     REM CALCULATE FACTOR OF SAFETY, INITIAL GUESS 1.0
330     REM
340     F=1.0
350     T=0.0
360     B=0.0
370     FOR I=1 TO N
380     G=1 + TAN(A(I))*TAN(P(I))/F
390     T=T + (C(I)*L(I) + TAN(P(I))*(W(I)/COS(A(I)) - U(I)*L(I)))/G
400     B=B + W(I)*SIN(A(I))
410     NEXT I
420     FL = F
430     F=T/B
440     IF ABS(FL/F-1)>0.0001 GOTO 350
450     PRINT "FACTOR OF SAFETY (SIMPLIFIED BISHOP METHOD) IS",F
460     REM
470     REM OPTIONS ON CHANGING DATA
480     REM
490     PRINT "ENTER -2    TO CHANGE POSITION OF SLIP SURFACE"
500     PRINT "      -1    TO CALCULATE FACTOR OF SAFETY"
510     PRINT "       0    TO EXIT"
520     PRINT "       1..N TO CHANGE DATA FOR A SLICE (GIVE SLICE NUMBER)"
530     INPUT I
540     IF I=0 THEN STOP
550     IF I=-1 GOTO 240
560     IF I=-2 GOTO 600
570     GOSUB 1000
580     GOSUB 2000
590     GOTO 490
600     PRINT "ENTER NEW YB COORDINATES ALONG SLIP SURFACE"
610     FOR I=0 TO N
620     INPUT YB(I)
630     NEXT I
640     GOTO 240
1000    PRINT "ENTER     C, PHI, GAMMA, X, YB, YT, YW"
1010    PRINT "ENTER KN/M2, DEG, KN/M3, M,  M,  M,  M"
1020    RETURN
2000    PRINT "SLICE NUMBER",I
```

```
2010     INPUT C(I),P(I),GA(I),X(I),YB(I),YT(I),YW(I)
2020     P(I) = P(I) * 3.14159 / 180.0
2030     RETURN
2040     END

READY

RUN

ENTER NUMBER OF SLICES
? 5
ENTER X COORDINATE OF TOP OF SLOPE (M):
? 4
ENTER BOTTOM AND TOP Y COORDS., LEFT SIDE OF TOP SLICE (M);
? 50, 50
ENTER WATER TABLE LEVEL AT TOP OF SLOPE (M):
? 0
ENTER DATA FOR EACH SLICE:
ENTER      C, PHI, GAMMA, X, YB, YT, YW
ENTER KN/M2, DEG, KN/M3, M,  M,  M,  M
SLICE NUMBER    1
? 25, 30, 21.6, 20, 27.2, 45, 0
SLICE NUMBER    2
? 25, 30, 21.6, 40, 12, 35, 0
SLICE NUMBER    3
? 25, 30, 21.6, 60, 3.5, 25, 0
SLICE NUMBER    4
? 25, 30, 21.6, 80, 0.5, 15, 0
SLICE NUMBER    5
? 25, 30, 21.6, 105, 2.8, 2.8, 0
FACTOR OF SAFETY (SIMPLIFIED BISHOP METHOD) IS          1.75374
ENTER -2     TO CHANGE POSITION OF SLIP SURFACE
      -1     TO CALCULATE FACTOR OF SAFETY
       0     TO EXIT
       1..N TO CHANGE DATA FOR A SLICE (GIVE SLICE NUMBER)
? -2
ENTER NEW YB COORDINATES ALONG SLIP SURFACE
? 50
? 25
? 9
? 1
? -1.5
? 2.8
FACTOR OF SAFETY (SIMPLIFIED BISHOP METHOD) IS          1.82573
ENTER -2     TO CHANGE POSITION OF SLIP SURFACE
      -1     TO CALCULATE FACTOR OF SAFETY
       0     TO EXIT
       1..N TO CHANGE DATA FOR A SLICE (GIVE SLICE NUMBER)
? 0
```

Program notes

(1) Although the program in Example 6.2 implements Bishop's method quite efficiently, it is not very convenient to use because of the form of input required. An engineer usually works by drawing a possible slip surface; for the programs in Examples 6.1 and 6.2 he then has to measure the values of α and l and calculate the values of W and u for each slice. Some of this work can be done by the computer. It is more convenient to enter the co-ordinates of the top and bottom of each slice, and the level of the water table. The computer then calculates the weight of the slice, the average water pressure on the base and the parameters α and l for the slice. This method is used in this example.

(2) Lines 10 to 200 input the slightly modified data. Note that information for each slice is requested by the subroutine in lines 2000 to 2040 since this is also used in the procedure to modify the parameters for a slice. Lines 210 to 300 are used to 'pre-process' the data from the convenient form of input to the form in which it is required for Bishop's calculation. This form of processing of input before a main calculation is often used.
(3) The calculation of the factor of safety is also made easier for the user. Making use of the information from Example 6.2 that only a few iterations will be necessary, the iterating process is made automatic. Line 440 checks if a sufficiently accurate value for the factor of safety has been calculated.
(4) Again various options are made available for changing the data. The most usual case is that the engineer will want to examine the effect of changing the position of the slip surface, so by entering −2 at line 530 the option is given to input a set of new YB values to analyse a new slip surface, as shown in Figure 6.7. This option is used in the sample run and the new failure surface proves to be slightly less critical than the first one.

Example 6.4 Calculation of inclinometer profiles

Write a program to analyse the results of an inclinometer profile. The measurements should be referred to a zero position either at the bottom of the tube (if it is fixed into a hard stratum) or the top of the tube (which may be located by surveying).

```
LIST

10      REM EXAMPLE 6.4. CALCULATION OF INCLINOMETER PROFILES
20      REM
30      REM BASE LENGTH FOR INCLINOMETER (M)
40      DATA 0.5
50      REM
60      DIM R(100),D(100)
70      READ BL
80      PRINT "REFER DISPLACEMENTS TO TOP (T) OR BOTTOM (B)";
90      INPUT T$
100     IF T$="T" GOTO 120
110     IF T$<>"B" GOTO 80
120     PRINT "ENTER CALIBRATION FACTOR (RADIANS/READING):"
130     INPUT F
140     N=0
150     PRINT "ENTER INCLINATION VALUES, STARTING AT BASE (-999 TO END):"
160     N=N+1
170     INPUT R(N)
180     IF R(N)<>-999 GOTO 160
190     N=N-1
200     D(0)=0
210     FOR I=1 TO N
220     D(I) = D(I-1) + BL * F * R(I)
230     NEXT I
240     IF T$="B" GOTO 280
250     FOR I=0 TO N
260     D(I) = D(I) - D(N)
270     NEXT I
280     REM
```

```
290       REM OUTPUT THE RESULTS
300       REM
310       PRINT " "
320       PRINT "HEIGHT ABOVE BASE","DISPLACEMENT"
330       PRINT "(M)"," ","(M)"
340       PRINT " "
350       FOR I=0 TO N
360       PRINT BL*I," ",D(I)
370       NEXT I
380       STOP
390       END

READY

RUN

REFER DISPLACEMENTS TO TOP (T) OR BOTTOM (B)? B
ENTER CALIBRATION FACTOR (RADIANS/READING):
? 0.01087
ENTER INCLINATION VALUES, STARTING AT BASE (-999 TO END):
? 0.584
? 0.846
? 0.585
? 0.583
? 1.275
? 1.489
? 1.867
? 0.21
? 0.51
? -0.678
? -2.89
? -5.893
? -8.35
? -9.87
? -999

HEIGHT ABOVE BASE              DISPLACEMENT
(M)                            (M)

 0                              0
 .5                             .317404E-02
 1                              .777205E-02
 1.5                            .109515E-01
 2                              .141201E-01
 2.5                            .210498E-01
 3                              .291425E-01
 3.5                            .392896E-01
 4                              .040431
 4.5                            .432028E-01
 5                              .395179E-01
 5.5                            .238107E-01
 6                             -.821772E-02
 6.5                           -.0536
 7                             -.107243
```

Program notes

(1) The program uses a string variable T$ to record whether the data is to be referred to the top or bottom of the tube.
(2) Each inclinometer may use a different calibration factor which must be entered by the programmer.
(3) The program first calculates the displacements referred to the bottom of the tube by summing the contributions from each 0.5 m length of sloping tube. If reference is required from the top then the necessary adjustment is made at lines 250 to 270.

PROBLEMS

(6.1) Many natural slopes may be approximated by the infinite slope considered in section 6.2. The stability of the slope is usually critically dependent on the nature of the groundwater flow within the slope. Write a program based on Equation (6.1), allowing each of the variables ϕ', α, β and γ to be altered for successive calculations, and use it to make a parametric study of the stability of a saturated slope.

(6.2) Extend the program from Problem (6.1) to include Equations (6.4) and (6.5) so that these may be selected if required for a particular case. Investigate the way in which the calculated factor of safety depends on the assumed depth to the failure plane.

(6.3) Write a program to determine the factor of safety of a slope in homogeneous clay using a total stress analysis of a circular slip as discussed in section 6.3. Start by considering the case of the infinite slope shown in Figure 6.3. The required input should be the distance of the centre of the circle from the slope, the radius of the circle, and the unit weight and undrained shear strength of the clay. Consider that the slipping body of soil makes a small rotational movement and compare the work done by gravity on the mass of soil with the work available to resist slipping by cohesion along the failure surface. The factor of safety should be as given in Equation (6.6).

The program can then be extended to consider the more difficult geometrical problems posed by circular failure planes of the types shown in Figure 6.4, using the same type of calculation. The accuracy of the results obtained may be checked against Taylor's charts.

(6.4) Modify the program in Example 6.3 so that it uses the factor of safety calculated by the Swedish method of slices as the starting value for the Bishop calculation. The program should print out the value of F obtained from both calculations.

(6.5) It is often the case that the most critical slip surface is almost circular. Write a program to carry out an analysis of a circular slip surface taking the following information as input:

(a) a set of co-ordinates giving the ground surface profile
(b) values for the strength and density of the soil
(c) the radius of the slip circle and the co-ordinates of its centre.

The program should divide the sliding region into a suitable number of slices and calculate the necessary quantities for each slice automatically. The geometrical calculations required for the program for the last part of Problem (6.3) should prove a useful starting point.

Begin by writing a simple program which uses the Swedish method of slices and does not include pore-water pressures. When this is working change a few lines to carry out the Bishop calculation, then add the effect of pore pressures by specifying also a profile of the water table.

(6.6) It is possible to search for the most critical slip surface by varying the radius of the slip circle and its central position. This is tedious if all the values are to be entered by hand, and it is possible to write a program to do some of the work automatically.

The most critical slip surface usually passes through the toe of a slope. Adapt one of the programs so that a set of circle centre co-ordinates is input at the start, the program then calculating the factor of safety for a 'toe-circle' with each centre in turn. In practice it is best to choose centres on a grid pattern approximately above the middle of the slope as shown in Figure 6.9. Once all the factors of safety have been calculated it is easy to identify the location of the most critical centre and the corresponding minimum factor of safety.

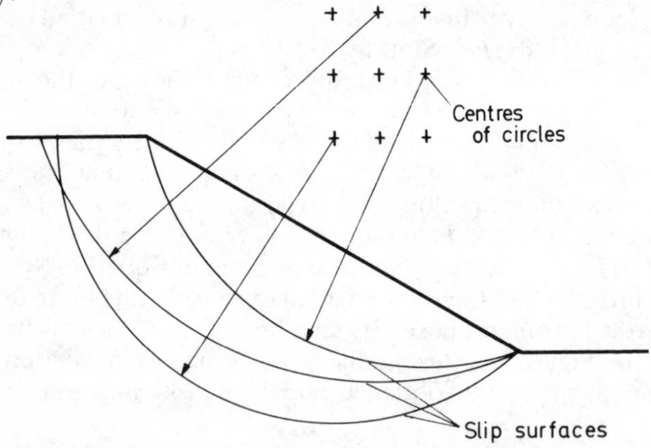

Figure 6.9 Typical grid of centres for critical slip circles

(6.7) (Challenge problem). The process of automation may be taken further by writing a program which searches for the most critical surface. The program should take as input an initial slip surface centre and while keeping the *x* co-ordinate constant should seek the most critical *y* co-ordinate for the centre, each slip surface passing through the toe of the slope. The method of search used in the Coulomb analyses in Examples 5.2 and 5.3 may be useful.

Since the program has to calculate many factors of safety, some attention should be paid to making this calculation as fast as possible. Once the search for the most critical *y* co-ordinate is successful, a similar search for the most critical *x* co-ordinate may be made as well.

Note that this is not an easy problem, and there is much to learn about the difficulties of numerical analysis by trying to solve it.

Chapter 7

Foundations

ESSENTIAL THEORY

7.1 Working and collapse conditions

Foundation problems include all those in which predominantly vertical loading is applied to the ground. Examples are the loading from buildings, whether through individual column footings or a continuous raft foundation, or from embankments or dams. There are usually two concerns, one or other of which may be more important in a particular situation. The first is that the ground has adequate bearing capacity; there must be a sufficient factor of safety against a complete collapse of the soil. This requirement is similar to that considered in the previous two chapters, and similar methods of analysis are applicable. The second is that movements of the ground, in particular vertical settlements, are not so great as to damage the structure or cause it to become unserviceable. The latter is concerned with working conditions, when the stresses in the ground are well below those which would cause failure of the soil, and analysis depends on the compressibility of the soil rather than its strength.

7.2 Bearing capacity of shallow footings

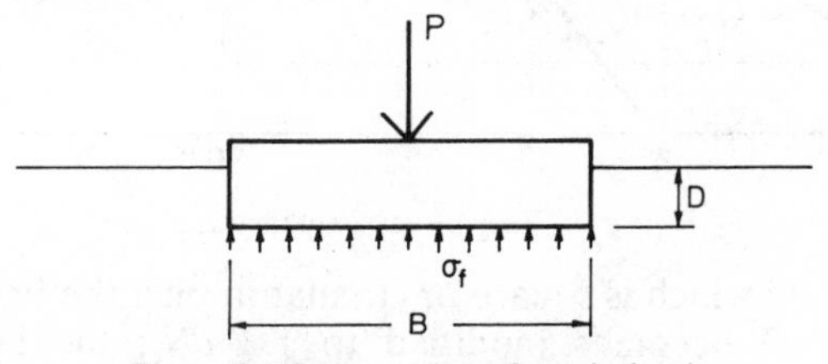

Figure 7.1 Cross-section of a strip footing

The ultimate load that may be applied to a strip footing resting on or near the surface of the ground (see Figure 7.1) may be determined for various idealized soil conditions by both upper and lower bound methods, which in some cases converge to give exact solutions. Such analyses consider independently the contributions to bearing capacity of cohesion, surcharge pressure on the soil surface, and the weight of the soil. Although the mode of failure is not necessarily the same in each case and the principle of superposition does not apply, Terzaghi argued that the total ultimate load could be obtained to a good approximation by adding the three contributions to give

$$\sigma_f = P/B = cN_c + \gamma DN_q + (\gamma B/2)N_\gamma \tag{7.1}$$

in which c is the cohesion of the soil and γD is the surcharge at the level of the base of the footing due to the depth of embedment of the footing. N_c, N_q and N_γ are bearing capacity factors determined from the analyses mentioned above; they are all functions of the friction angle for the soil as shown in Figure 7.2. Equation (7.1) may be applied either to 'undrained' conditions (for which $c = c_u$ and $\phi = 0$ normally) or 'drained' conditions (for which $\phi = \phi'$ and $c = 0$ normally), using appropriate values for γ for the soil above and below the level of the footing in the second and third term respectively. The undrained analysis will give σ_f as a total stress; the drained analysis will give σ_f as an effective stress, to which any pore pressure on the base of the footing must be added to give the total stress.

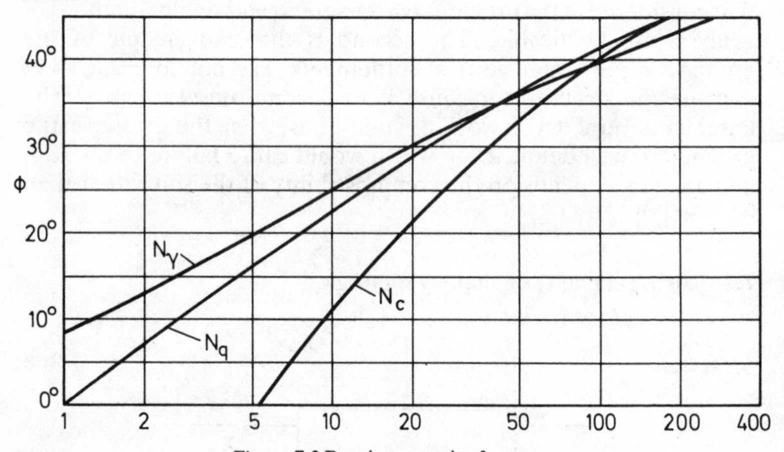

Figure 7.2 Bearing capacity factors

For a footing which is square or circular in plan the first term in Equation (7.1) becomes modified to 1.2 cN_c; the third term becomes $0.4\gamma B$ for a square footing, and $0.6\gamma R$ for a circular footing of radius R.

When the load P is eccentric or inclined to the vertical the ultimate load may be substantially reduced. Meyerhof has suggested that loading of eccentricity e may be taken into account by using an effective width B', where

$$B' = B - 2e \tag{7.2}$$

and that loading with inclination α may be taken into account by multiplying N_c, N_q and N_γ by inclination factors i_c, i_q and i_γ where

$$i_c = i_q = (1 - 2\alpha/\pi)^2 \tag{7.3}$$

and

$$i_\gamma = (1 - \alpha/\phi)^2 \tag{7.4}$$

where both α and ϕ are expressed in radians.

7.3 Foundations in clay

Foundations generally involve an increase in stresses on the ground, causing the soil to consolidate and increase in strength with time. For clays the short term, undrained conditions are therefore usually the most critical. Equation (7.1) then reduces to

$$\sigma_f = N_c c_u + \gamma D \tag{7.5}$$

since $N_q = 1$ and $N_\gamma = 0$. N_c has the value 5.14 if the base of the footing rests on the surface of the soil and $\phi = 0$. For a circular footing the equivalent values are 5.69 for a smooth footing and 6.04 if the base of the footing is rough. Although it is usual to ignore the strength of any soil above the base of a footing, treating it simply as a surcharge, it is common to use higher N_c values for embedded footings with $\phi = 0$. Values of 7.5 for deep strip foundations and 9.0 for deep circular foundations are often used. The latter value is appropriate for the 'end bearing' capacity of piles. For intermediate depths of embedment and rectangular footings of dimensions $B \times L$ ($B < L$) the value of N_c is given approximately by:

$$N_c = 5(1 + 0.2D/B)(1 + 0.2B/L) \tag{7.6}$$

with a maximum value of 1.5 for the term in the first bracket, so that no further increase is given for $D > 2.5B$.

Deep foundations such as piles and caissons transfer load to the ground by friction or adhesion at the vertical sides of the foundation as well as bearing pressure on the base (Figure 7.3). Under undrained conditions the adhesion is taken to be αc_u; α is typically about 1.0 for soft clays, decreasing to 0.3 for very heavily overconsolidated clays, with a typical value for stiff clays of 0.45. In some cases the ground around a pile settles after installation of the pile and induces a downdrag or 'negative skin friction' loading on the pile, as shown in Figure 7.3.

Piles in groups tend to interfere with each other so that the ultimate load capacity of each pile is reduced by an efficiency factor which may typically be about 0.7 for pile spacings of 2 to 3 dia-

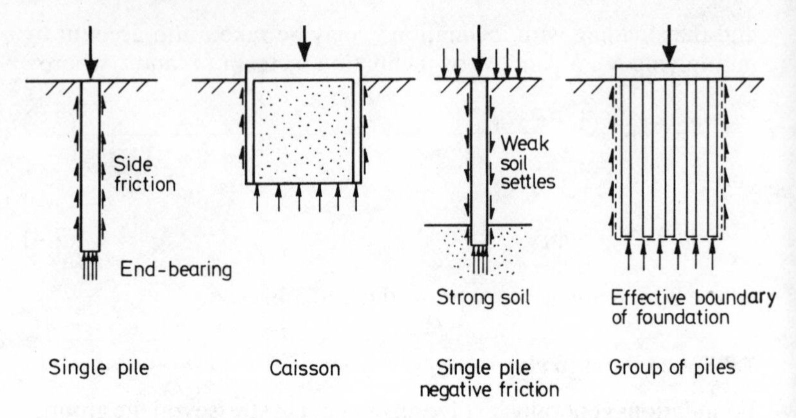

Figure 7.3 Types of deep foundation

meters. However if the piles are connected by a pilecap or raft foundation resting on the soil, failure can only occur by a block failure in which the whole group must be considered as a single foundation, as shown in Figure 7.3.

All the above methods of analysis, except for calculation of side friction on piles, assume reasonably homogeneous soil conditions. In practice it is often possible to treat the soil as approximately homogeneous under footings of limited size since the failure occurs within a depth less or not much greater than the width of the footing. However when the soil contains layers with markedly different properties, or when the strength decreases rather than increases with depth, it may be necessary to resort to upper bound or limit-equilibrium type analyses, varying the failure surface to determine the most critical conditions. The methods of calculation are similar to those introduced in the previous two chapters and will not be covered further in this chapter.

7.4 Elastic analysis of foundations

When soil properties are reasonably constant to a depth of more than about five times the width of a footing the soil may be considered to behave approximately as a bed of homogeneous, isotropic, linearly elastic material of infinite depth, provided the stresses are nowhere sufficient to cause yield or failure in the soil. Standard solutions from the theory of elasticity may then be applied to determine the stresses, strains and displacements caused by the loading on the foundation. A useful compilation of such solutions has been made by Poulos and Davis (Reference 10). Many are presented in the form of charts or tables, but some closed-form solutions are available, are suitable for incorporation in computer programs, and are given below. Figure 7.4 shows two types of surface loading on a strip of width B. For the uniform pressure the stresses on a soil element at (x, z) are given by

$$\sigma_z = \frac{\sigma_o}{\pi}\{\alpha + \sin\alpha\cos(\alpha + 2\beta)\} \tag{7.7}$$

$$\sigma_x = \frac{\sigma_o}{\pi}\{\alpha - \sin\alpha\cos(\alpha + 2\beta)\} \tag{7.8}$$

$$\tau_{xz} = \frac{\sigma_o}{\pi}\{\sin\alpha\sin(\alpha + 2\beta)\} \tag{7.9}$$

For the linearly increasing pressure the stresses are given by

$$\sigma_z = \frac{\sigma_o}{\pi}\{\frac{x}{B}\alpha - \tfrac{1}{2}\sin 2\beta\} \tag{7.10}$$

$$\sigma_x = \frac{\sigma_o}{\pi}\{\frac{x}{B}\alpha - \frac{z}{B}\ln(R_1^2/R_2^2) + \tfrac{1}{2}\sin 2\beta\} \tag{7.11}$$

$$\tau_{xz} = \frac{\sigma_o}{\pi}\{\tfrac{1}{2} + \tfrac{1}{2}\cos 2\beta - z\alpha/B\} \tag{7.12}$$

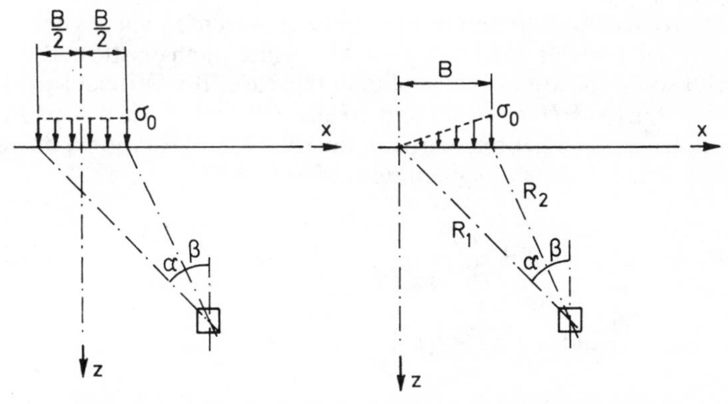

Figure 7.4 Soil elements beneath uniformly loaded strip footings

For both the above cases absolute values for the surface settlements are indeterminate, although it is possible to relate the displacements at two points provided neither is at infinity.

The principle of superposition applies to such elastic solutions so that the effect of various types of loading, such as from an embankment or a spoil heap, may be determined by approximating the loading to a combination of uniform and linearly increasing pressures on the surface and combining the resulting stresses and deflections. The magnitudes of the principal stresses and of mean normal stress p and shear stress q may be determined from the stresses calculated from Equations (7.7) to (7.9) and (7.10) to (7.12) using the usual relationships of stress analysis.

For instance, under the uniform strip loading the principal

stresses are given by

$$\sigma_{1,3} = \frac{\sigma_o}{\pi}[\alpha \pm \sin\alpha] \tag{7.13}$$

and the maximum shear stress at any point by

$$\tau = \frac{\sigma_o}{\pi}\sin\alpha \tag{7.14}$$

In Equation (7.14) the maximum value of τ occurs when α is 90°. Maximum shear stresses therefore occur at points lying on a semi-circle with the strip footing as diameter, and have magnitude σ_o/π. For a surface on clay with undrained shear strength c_u it is therefore predicted that local yielding of the soil under the footing, leading to rapidly increasing settlements, will occur when the loading on the footing is πc_u; complete plastic failure of the soil will not occur until the loading reaches $(\pi + 2)c_u$ in accordance with Equation (7.5).

Many structural foundations may be represented approximately by a rough circular pad resting on the surface of an elastic bed, and useful solutions are also available in this case. If a vertical load V, horizontal load H and moment M are applied to such a pad of diameter B, the resulting vertical, horizontal and rotational movements of the footing are given respectively by

$$\delta_v = \frac{V}{EB}(1 - \nu^2) \tag{7.15}$$

$$\delta_h = \frac{H}{2EB}(1 + \nu)(2 - \nu) \tag{7.16}$$

$$\theta = \frac{6M}{EB^3}(1 - \nu^2) \tag{7.17}$$

The pressure on the base of such a pad will not generally be uniform, but in practice may often be taken to be approximately so; in some cases, such as under an oil storage tank with a circular base, the contact pressure will in fact be almost uniform. Under the centre of a circular area of diameter B carrying a uniform vertical loading of σ_o the vertical, radial and circumferential stresses (which are principal stresses) are given by

$$\sigma_z = \sigma_o\left[1 - \left\{\frac{1}{1 + \left(\frac{B}{2z}\right)^2}\right\}^{\frac{3}{2}}\right] \tag{7.18}$$

$$\sigma_r = \sigma_\theta = \frac{\sigma_o}{2}\left[(1+2\nu) - \frac{2(1+\nu)}{\left\{1+\left(\frac{B}{2z}\right)\right\}^{\frac{1}{2}}} + \frac{1}{\left\{1+\left(\frac{B}{2z}\right)\right\}^{\frac{3}{2}}}\right] \quad (7.19)$$

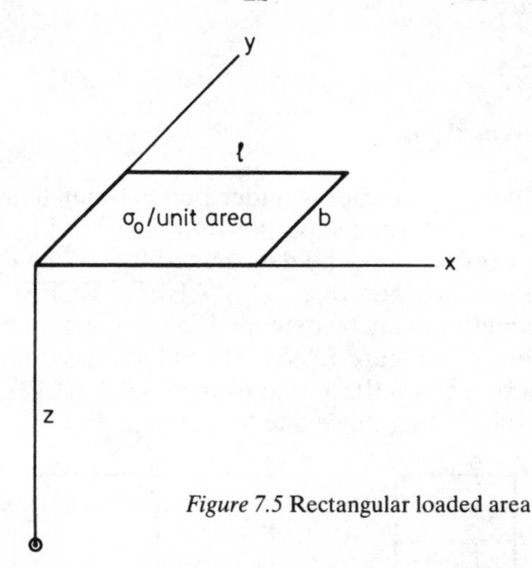

Figure 7.5 Rectangular loaded area

If the loaded area is rectangular, as shown in Figure 7.5, expressions are available for stresses and deflections beneath a corner of such an area carrying a uniform loading of σ_o. The vertical stress and displacement respectively at a depth z are given by

$$\sigma_z = \frac{\sigma_o}{2\pi}\left[\tan^{-1}\frac{lb}{zR_3} + \frac{lbz}{R_3}\left(\frac{1}{R_1^2} + \frac{1}{R_2^2}\right)\right] \quad (7.20)$$

where

$$R_1 = (l^2 + z^2)^{\frac{1}{2}}$$

$$R_2 = (b^2 + z^2)^{\frac{1}{2}}$$

$$R_3 = (l^2 + b^2 + z^2)^{\frac{1}{2}}$$

and

$$\rho_z = \frac{\sigma_o b}{E}(1-\nu^2)\left[A - \frac{(1-2\nu)}{(1-\nu)}B\right] \quad (7.21)$$

where

$$A = \frac{1}{2\pi}\left[\ln\frac{(c+m_1)}{(c-m_1)} + m_1 \ln\frac{(c+1)}{(c-1)}\right]$$

$$B = \frac{n_1}{2\pi}\tan^{-1}\frac{m_1}{n_1 c}$$

$$m_1 = l/b$$

$$n_1 = z/b$$

and $c = (1 + m_1^2 + n_1^2)^{\frac{1}{2}}$

Stresses and displacements under points other than the corners may be determined using superposition. Thus in Figure 7.6(a) the values under point E may be determined by adding the values for the corners of the four rectangles ABED, BCFE, DEHG and EFJH. The method may be extended to more complex rectangular shapes as shown in Figure 7.6(b). The values under point P may be determined by adding those due to rectangles KLPN, LMRP and PRUS and subtracting those due to rectangle PQTS.

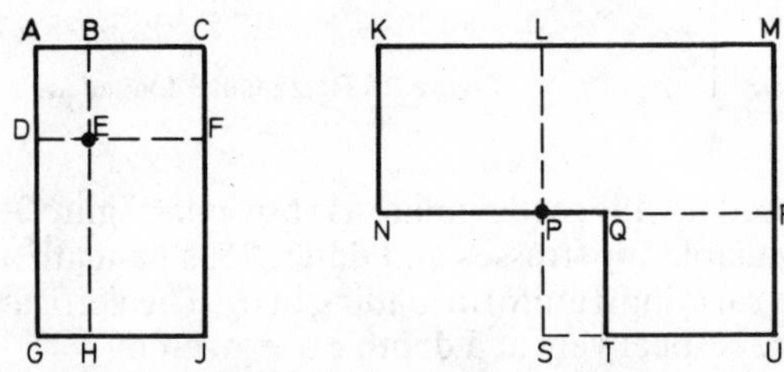

Figure 7.6 Superposition of rectangular loaded areas

Where foundation shapes or soil conditions are more complex solutions may still be obtained using influence factors, charts and tables which may be found in Reference 10 and many standard texts on soil mechanics. When the size of the project warrants it, a numerical solution by finite element methods may be undertaken. Neither of these approaches will be considered further in this book.

However, many of the algebraic expressions in Equations (7.7) to (7.21) are quite complex. Evaluation of them by hand is both tedious and likely to lead to errors. The use of computer programs to perform this task is valuable in avoiding both problems; four of the example programs in this Chapter are therefore concerned with analyses of this type.

In all the equations for elastic settlements the values of E and ν may be chosen as the undrained values, giving the settlements occurring immediately on loading, or the drained values, giving the final long term settlements after pore pressures have returned to equilibrium conditions. The difference between the two settlements is the amount of settlement that occurs due to consolidation; this is of importance with clay soils and is considered further in the next section.

7.5 Consolidation settlements of foundations on clay

In practice the drained elastic parameters E' and ν' are not often obtained. Settlements of foundations on clay are calculated in two stages, the first to determine the immediate settlements using an elastic analysis (with $\nu_u = 0.5$, E_u often obtained from empirical correlations with c_u), the second to obtain the settlements due to consolidation.

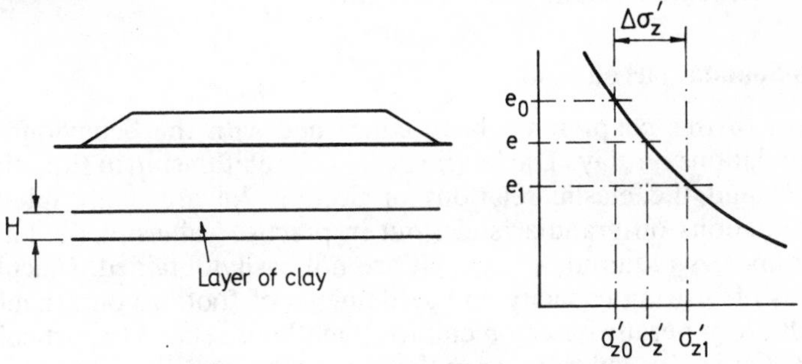

Figure 7.7 Consolidation of a clay layer beneath an embankment

A thin layer of clay at some depth below a footing or embankment (see Figure 7.7) will experience an increase in vertical stress due to the applied loading which may be estimated from equations such as (7.7), (7.10), (7.18) or (7.20). If the area over which the load is spread is reasonably large compared with the thickness of the layer of clay its consolidation behaviour will be similar to, and may be modelled by, that of a specimen of the clay in an oedometer test (see section 4.3) undergoing one dimensional compression. The compression of the layer may then be calculated either from the initial and final void ratios at the corresponding stress levels (see Figure 7.7) or in terms of the coefficient of volume change m_v determined from the oedometer test.

$$\Delta H = \frac{(e_0 - e_1)H}{(1 + e_0)} = m_v . \Delta\sigma'_z . H \tag{7.22}$$

where

$$m_v = \frac{C_c}{\sigma_z'(1+e)} \tag{7.23}$$

In Equation (7.23) C_c is the compression index (see Equation 4.9), σ'_z the mean effective vertical stress in the clay, and e the mean voids ratio corresponding to σ'_z. Since C_c is a constant, m_v clearly varies with stress level and must be calculated for appropriate stresses.

The rate at which settlement occurs may also be calculated using the one dimensional analysis discussed in section 4.4. Such calculations may be of considerable importance, for instance in deciding at what stage a relatively inflexible road or other structure may be built on top of an embankment without fear of damage due to settlement still to take place.

For thicker layers of clay the one dimensional approach may still be used by dividing the stratum up into horizontal layers for each of which the compression is calculated using Equation (7.22) and the total settlement obtained by summation.

7.6 Foundations on sand

Most of this chapter has been concerned with the behaviour of foundations on clay. The bearing capacity relationship in Equation (7.1) and the elastic relations of Section 7.4 are applicable to foundations on granular soils, but in practice values of the basic parameters required, ϕ', E', ν', are not easily obtained. Calculations of bearing capacity and settlements of footings on granular soils are generally based on empirical methods related to particular types of in situ soil tests, or in the case of driven piles to the actual behaviour of the pile recorded during driving. Such calculations are not really appropriate for this book, but are covered in standard texts on soil mechanics and foundation engineering.

WORKED EXAMPLES

Example 7.1 Calculation of bearing capacity of a strip footing

Write a program to determine the bearing capacity of a strip footing using Terzaghi's method of bearing capacity factors (Equation (7.1)). The bearing capacity factors may be determined from

$$N_q = \tan^2\left(\frac{\pi}{4}+\frac{\phi}{2}\right)e^{\pi\tan\phi} \tag{7.24}$$

$$N_c = (N_q - 1)/\tan\phi \tag{7.25}$$

$$N_\gamma = 1.5(N_q - 1)\tan\phi \tag{7.26}$$

The first two give exact theoretical values, the last is an approximation which fits the theoretical value quite closely.

```
LIST

10      REM EXAMPLE 7.1. BEARING CAPACITY CALC. USING APPROX. FACTORS
20      REM
30      PRINT "ENTER PROPERTIES FOR THE SOIL:"
40      PRINT "C (KN/M2)";
50      INPUT C
60      PRINT "PHI (DEG)";
70      INPUT PH
```

```
80        PH = PH * 3.14159 / 180.0
90        PRINT "GAMMA (KN/M3)";
100       INPUT GA
110       NQ = EXP(3.14159 * TAN(PH)) * ((TAN(3.14159*0.25 + PH*0.5)) ^ 2)
120       NG = 1.5 * (NQ - 1) * TAN(PH)
130       IF PH=0 GOTO 160
140       NC = (NQ - 1) / TAN(PH)
150       GOTO 170
160       NC = 5.14159
170       PRINT "BEARING CAPACITY FACTORS ARE:"
180       PRINT "NC",NC
190       PRINT "NQ",NQ
200       PRINT "NGAMMA",NG
210       PRINT "ENTER DIMENSIONS OF FOUNDATION:"
220       PRINT "B (M)";
230       INPUT B
240       PRINT "D (M)";
250       INPUT D
260       P = B * (C*NC + GA*D*NQ + 0.5*B*GA*NG)
270       PRINT "CAPACITY OF FOOTING IS: ",P;" KN PER METRE LENGTH"
280       PRINT "ENTER 1 TO STOP"
290       PRINT "      2 TO CHANGE SOIL PROPERTIES"
300       PRINT "      3 TO CHANGE FOOTING DIMENSIONS"
310       INPUT I
320       ON I GOTO 330,30,210
330       STOP
340       END

READY

RUN

ENTER PROPERTIES FOR THE SOIL:
C (KN/M2)? 128
PHI (DEG)? 0
GAMMA (KN/M3)? 19.7
BEARING CAPACITY FACTORS ARE:
NC             5.14159
NQ             .999997
NGAMMA         0
ENTER DIMENSIONS OF FOUNDATION:
B (M)? 2.5
D (M)? 0
CAPACITY OF FOOTING IS:     1645.31  KN PER METRE LENGTH
ENTER 1 TO STOP
      2 TO CHANGE SOIL PROPERTIES
      3 TO CHANGE FOOTING DIMENSIONS
? 3
ENTER DIMENSIONS OF FOUNDATION:
B (M)? 2.5
D (M)? .5
CAPACITY OF FOOTING IS:     1669.93  KN PER METRE LENGTH
ENTER 1 TO STOP
      2 TO CHANGE SOIL PROPERTIES
      3 TO CHANGE FOOTING DIMENSIONS
? 2
ENTER PROPERTIES FOR THE SOIL:
C (KN/M2)? 0
PHI (DEG)? 37
GAMMA (KN/M3)? 9.7
BEARING CAPACITY FACTORS ARE:
NC             55.6292
NQ             42.9195
NGAMMA         47.3829
ENTER DIMENSIONS OF FOUNDATION:
B (M)? 2.5
D (M)? 0
CAPACITY OF FOOTING IS:     1436.29  KN PER METRE LENGTH
```

```
ENTER 1 TO STOP
      2 TO CHANGE SOIL PROPERTIES
      3 TO CHANGE FOOTING DIMENSIONS
? 3
ENTER DIMENSIONS OF FOUNDATION:
B (M)? 2.5
D (M)? .5
CAPACITY OF FOOTING IS:     1956.69  KN PER METRE LENGTH
ENTER 1 TO STOP
      2 TO CHANGE SOIL PROPERTIES
      3 TO CHANGE FOOTING DIMENSIONS
? 1
```

Program notes

(1) First the soil properties, c, ϕ and γ are entered and the bearing capacity factors calculated from Equations (7.24) to (7.26) as a function of ϕ. Note that the special case of $\phi = 0$ must be dealt with separately for N_c as Equation (7.25) is indeterminate for this value.
(2) For any set of values of breadth and depth of footing the program then calculates the bearing capacity per unit length of footing. The sample runs show a simple case of total and effective stress analysis of a footing at two different depths.

Example 7.2 Stresses under an uniformly loaded circular area

In planning oedometer tests to examine the compressibility of soils, and in subsequently calculating consolidation settlements, it is necessary to know the initial stresses in the ground and the increase in vertical stress due to any subsequent loading. Write a program to calculate the increase in stress beneath the centre of a uniformly loaded circular area; such an idealization would be applicable for instance to the stresses beneath an oil tank.

```
LIST

10     REM EXAMPLE 7.2. CALCULATION OF STRESS UNDER CENTRELINE OF
20     REM UNIFORMLY LOADED AREA
30     REM
40     PRINT "ENTER RADIUS OF LOADED AREA (M):"
50     INPUT A
60     PRINT "ENTER INTENSITY OF LOADING (KN/M2):"
70 .   INPUT Q
80     PRINT "ENTER MAXIMUM DEPTH FOR CALCULATION OF STRESS (M):"
90     INPUT ZM
100    PRINT "ENTER NUMBER OF STRESS POINTS TO BE CALCULATED:"
110    INPUT N
120    PRINT "ENTER INSITU BULK UNIT WEIGHT (KN/M3):"
130    INPUT GA
140    PRINT "ENTER DEPTH TO GROUND WATER TABLE (M):"
150    INPUT W
160    REM
170    REM CALCULATE INSITU STRESS, INCREMENT AND FINAL VALUE
180    REM
190    PRINT " "
200    PRINT "DEPTH (M)","VERTICAL EFFECTIVE STRESS (KN/M2)"
210    PRINT " ","INITIAL","INCREMENT","FINAL"
220    PRINT " "
230    PRINT 0,0,Q,Q
```

```
240       FOR I=1 TO N
250       Z = ZM * I / N
260       ST = Z * GA
270       U = 9.81 * (Z - W)
280       IF U<0 THEN U=0
290       SE = ST - U
300       S = Q * (1 - 1 / ((1 + (A*A)/(Z*Z)) ^ 1.5))
310       PRINT Z,SE,S,(SE+S)
320       NEXT I
330       PRINT " "
340       PRINT "ENTER 0 TO END, 1 FOR NEW SET OF DATA"
350       INPUT J
360       IF J=1 GOTO 40
370       STOP
380       END

READY

RUN

ENTER RADIUS OF LOADED AREA (M):
? 5.5
ENTER INTENSITY OF LOADING (KN/M2):
? 73
ENTER MAXIMUM DEPTH FOR CALCULATION OF STRESS (M):
? 25
ENTER NUMBER OF STRESS POINTS TO BE CALCULATED:
? 10
ENTER INSITU BULK UNIT WEIGHT (KN/M3):
? 18.4
ENTER DEPTH TO GROUND WATER TABLE (M):
? 4.2

DEPTH (M)       VERTICAL EFFECTIVE STRESS (KN/M2)
                INITIAL        INCREMENT      FINAL

 0               0              73             73
 2.5             46             67.8275        113.827
 5               84.152         50.7805        134.932
 7.5             105.627        34.7191        140.346
 10              127.102        23.8915        150.994
 12.5            148.577        17.0198        165.597
 15              170.052        12.5846        182.637
 17.5            191.527        9.61895        201.146
 20              213.002        7.56178        220.564
 22.5            234.477        6.08618        240.563
 25              255.952        4.99633        260.948

ENTER 0 TO END, 1 FOR NEW SET OF DATA
? 0
```

Program notes

(1) The initial vertical effective stresses are calculated from the unit weight of the soil and the depth to the water table.
(2) Increments in stress are calculated from Equation (7.18), but note that in the program the load intensity is called Q and the loaded area is defined by its radius A rather than its diameter. Although elasticity theory was used to derive the expression the increase in vertical stress is independent of the values of E and ν.

Example 7.3 Stresses under a strip footing

Write a program to determine all the components of stress at any point under a strip footing, using Equations (7.7) to (7.9), (7.13) and (7.14).

```
LIST

10      REM EXAMPLE 7.3. CALCULATION OF STRESSES UNDER A STRIP FOOTING
20      REM
30      PRINT "ENTER DETAILS OF FOOTING"
40      PRINT "WIDTH OF FOOTING (M):"
50      INPUT B
60      PRINT "INTENSITY OF LOAD ON FOOTING (KN/M2):"
70      INPUT Q
80      PRINT "ENTER COORDINATES FOR STRESS CALCULATION"
90      PRINT "DEPTH (M):"
100     INPUT Z
110     PRINT "OFFSET FROM CENTRELINE (M):"
120     INPUT X
130     REM
140     REM CALCULATE INTERMEDIATE QUANTITIES AND PRINCIPAL STRESSES
150     REM
160     AL = ATN((X - 0.5*B)/Z)
170     BE = ATN((X + 0.5*B)/Z) - AL
180     S1 = Q * (BE + SIN(BE)) / 3.14159
190     S3 = Q * (BE - SIN(BE)) / 3.14159
200     TM = Q * SIN(BE) / 3.14159
210     REM
220     REM CALCULATE CARTESIAN STRESSES
230     REM
240     SX = Q * (BE - SIN(BE) * COS(BE + 2*AL)) / 3.14159
250     SZ = Q * (BE + SIN(BE) * COS(BE + 2*AL)) / 3.14159
260     T = Q * SIN(BE) * SIN(BE + 2*AL) / 3.14159
270     PRINT " "
280     PRINT "COORDINATES (M)"
290     PRINT "X",X
300     PRINT "Z",Z
310     PRINT "PRINCIPAL STRESSES (KN/M2)"
320     PRINT "SIG-1",S1
330     PRINT "SIG-3",S3
340     PRINT "TAU-MAX",TM
350     PRINT "CARTESIAN STRESSES (KN/M2)"
360     PRINT "SIG-X",SX
370     PRINT "SIG-Z",SZ
380     PRINT "TAU-XZ",T
390     PRINT " "
400     PRINT "ENTER 1 TO STOP"
410     PRINT "      2 FOR NEW POINT"
420     PRINT "      3 FOR NEW LOADS"
430     INPUT I
440     ON I GOTO 450,80,30
450     STOP
460     END

READY

RUN

ENTER DETAILS OF FOOTING
WIDTH OF FOOTING (M):
? 3.78
INTENSITY OF LOAD ON FOOTING (KN/M2):
? 58.2
ENTER COORDINATES FOR STRESS CALCULATION
DEPTH (M):
? 4.5
OFFSET FROM CENTRELINE (M):
? 0

COORDINATES (M)
```

```
X              0
Z              4.5
PRINCIPAL STRESSES (KN/M2)
SIG-1          27.9607
SIG-3          1.50453
TAU-MAX        13.2281
CARTESIAN STRESSES (KN/M2)
SIG-X          1.50453
SIG-Z          27.9607
TAU-XZ         0

ENTER 1 TO STOP
      2 FOR NEW POINT
      3 FOR NEW LOADS
? 2
ENTER COORDINATES FOR STRESS CALCULATION
DEPTH (M):
? 4.5
OFFSET FROM CENTRELINE (M):
? 1.86

COORDINATES (M)
X              1.86
Z              4.5
PRINCIPAL STRESSES (KN/M2)
SIG-1          24.9484
SIG-3          1.03954
TAU-MAX        11.9544
CARTESIAN STRESSES (KN/M2)
SIG-X          3.75951
SIG-Z          22.2285
TAU-XZ         7.59166

ENTER 1 TO STOP
      2 FOR NEW POINT
      3 FOR NEW LOADS
? 3
ENTER DETAILS OF FOOTING
WIDTH OF FOOTING (M):
? 1.86
INTENSITY OF LOAD ON FOOTING (KN/M2):
? 116.4
ENTER COORDINATES FOR STRESS CALCULATION
DEPTH (M):
? 4.5
OFFSET FROM CENTRELINE (M):
? 0

COORDINATES (M)
X              0
Z              4.5
PRINCIPAL STRESSES (KN/M2)
SIG-1          29.7892
SIG-3          .414697
TAU-MAX        14.6872
CARTESIAN STRESSES (KN/M2)
SIG-X          .414697
SIG-Z          29.7892
TAU-XZ         0

ENTER 1 TO STOP
      2 FOR NEW POINT
      3 FOR NEW LOADS
? 2
ENTER COORDINATES FOR STRESS CALCULATION
DEPTH (M):
? 4.5
OFFSET FROM CENTRELINE (M):
? 1.86
```

```
COORDINATES (M)
X               1.86
Z               4.5
PRINCIPAL STRESSES (KN/M2)
SIG-1           25.7589
SIG-3           .265846
TAU-MAX         12.7465
CARTESIAN STRESSES (KN/M2)
SIG-X           3.76264
SIG-Z           22.2621
TAU-XZ          8.77018

ENTER 1 TO STOP
      2 FOR NEW POINT
      3 FOR NEW LOADS
? 1
```

Program notes

(1) The program allows calculation of both the principal stresses and the stresses on the *x* and *y* planes at any depth and offset from the footing centreline. The program makes use of the angles α and β shown in Figure 7.4.
(2) The program is very inefficient in that there is a lot of repeated calculation in lines 160 to 200 and 240 to 260. This has been done to make the expressions for the stresses clear. If speed was of greater importance then these lines could be rewritten.
(3) The application given of the program in Example 7.3 is to examine the stresses which would be expected under a footing of given dimensions and how these change if the same load is applied to a footing of half the width.

Example 7.4 Elastic displacement of a rough circular footing

Write a program to calculate the displacements of a rough circular footing acted on by vertical, horizontal and moment loads.

```
LIST

10      REM EXAMPLE 7.4. ELASTIC DISPLACEMENTS OF A ROUGH CIRCULAR FOOTING
20      REM USES MENU OF COMMANDS PROMPTED BY "COMMAND?"
30      REM
40      PRINT "ELASTIC ANALYSIS OF CIRCULAR FOOTINGS"
50      GOSUB 8000
60      PRINT "COMMAND";
70      INPUT C$
80      IF C$="IF" THEN GOSUB 1000
90      IF C$="IS" THEN GOSUB 2000
100     IF C$="IL" THEN GOSUB 3000
110     IF C$="ID" THEN GOSUB 4000
120     IF C$="CS" THEN GOSUB 5000
130     IF C$="CD" THEN GOSUB 6000
140     IF C$="CL" THEN GOSUB 7000
150     IF C$="HE" THEN GOSUB 8000
160     IF C$="EX" THEN STOP
170     GOTO 60
1000    REM
1010    REM COMMAND IF - INPUT FOOTING RADIUS
1020    REM
```

```
1030     PRINT "ENTER RADIUS OF FOOTING (M):"
1040     INPUT R
1050     RETURN
2000     REM
2010     REM COMMAND IS - INPUT SOIL PROPERTIES
2020     REM
2030     PRINT "ENTER YOUNGS MODULUS FOR SOIL (KN/M2):"
2040     INPUT E
2050     PRINT "ENTER POISSONS RATIO FOR SOIL:"
2060     INPUT NU
2070     RETURN
3000     REM
3010     REM COMMAND IL - INPUT LOADS ON FOOTING
3020     REM
3030     PRINT "ENTER VERTICAL LOAD ON FOOTING (KN):"
3040     INPUT V
3050     PRINT "ENTER HORIZONTAL LOAD ON FOOTING (KN):"
3060     INPUT H
3070     PRINT "ENTER MOMENT ON FOOTING (KNM):"
3080     INPUT M
3090     RETURN
4000     REM
4010     REM COMMAND IL - INPUT DISPLACEMENTS OF FOOTING
4020     REM
4030     PRINT "ENTER VERTICAL DISPLACEMENT OF FOOTING (M):"
4040     INPUT DV
4050     PRINT "ENTER HORIZONTAL DISPLACEMENT OF FOOTING (M):"
4060     INPUT DH
4070     PRINT "ENTER ROTATION OF FOOTING (RADIANS):"
4080     INPUT TH
4090     RETURN
5000     REM
5010     REM COMMAND CS - CALCULATE STIFFNESS FACTORS
5020     REM
5030     F1 = 2*R*E / (1-NU^2)
5040     F2 = 4*R*E / ((1+NU)*(2-NU))
5050     F3 = 4*(R^3)*E / (3*(1-NU^2))
5060     PRINT "STIFFNESS FACTORS CALCULATED"
5070     RETURN
6000     REM
6010     REM COMMAND CD - CALCULATE DISPLACEMENTS
6020     REM
6030     DV = V / F1
6040     DH = H / F2
6050     TH = M / F3
6060     PRINT "VERTICAL DISPLACEMENT =   ",DV;" M"
6070     PRINT "HORIZONTAL DISPLACEMENT = ",DH;" M"
6080     PRINT "ROTATION =                ",TH;" RADIANS"
6090     RETURN
7000     REM
7010     REM COMMAND CL - CALCULATE LOADS
7020     REM
7030     V = DV * F1
7040     H = DH * F2
7050     M = TH * F3
7060     PRINT "VERTICAL LOAD =   ",V;" KN"
7070     PRINT "HORIZONTAL LOAD = ",H;" KN"
7080     PRINT "MOMENT =          ",M;" KNM"
7090     RETURN
8000     REM
8010     REM HELP COMMAND - SHOW COMMANDS AVAILABLE
8020     REM
8030     PRINT " "
8040     PRINT "COMMANDS AVAILABLE ARE:"
8050     PRINT " "
8060     PRINT "IF:  INPUT FOOTING DIMENSIONS"
8070     PRINT "IS:  INPUT SOIL PROPERTIES"
8080     PRINT "IL:  INPUT LOADS ON THE FOOTING"
```

```
8090     PRINT "ID:  INPUT DISPLACEMENTS OF FOOTING"
8100     PRINT "CS:  CALCULATE STIFFNESS FACTORS"
8110     PRINT "CD:  CALCULATE DISPLACEMENTS"
8120     PRINT "CL:  CALCULATE LOADS"
8130     PRINT "HE:  PRINT THIS HELP INFORMATION"
8140     PRINT "EX:  EXIT"
8150     PRINT " "
8160     RETURN
9000     END

READY

RUN

ELASTIC ANALYSIS OF CIRCULAR FOOTINGS

COMMANDS AVAILABLE ARE:

IF:  INPUT FOOTING DIMENSIONS
IS:  INPUT SOIL PROPERTIES
IL:  INPUT LOADS ON THE FOOTING
ID:  INPUT DISPLACEMENTS OF FOOTING
CS:  CALCULATE STIFFNESS FACTORS
CD:  CALCULATE DISPLACEMENTS
CL:  CALCULATE LOADS
HE:  PRINT THIS HELP INFORMATION
EX:  EXIT

COMMAND? IS
ENTER YOUNGS MODULUS FOR SOIL (KN/M2):
? 3450
ENTER POISSONS RATIO FOR SOIL:
? 0.3
COMMAND? IF
ENTER RADIUS OF FOOTING (M):
? 1.86
COMMAND? IL
ENTER VERTICAL LOAD ON FOOTING (KN):
? 2500
ENTER HORIZONTAL LOAD ON FOOTING (KN):
? 400
ENTER MOMENT ON FOOTING (KNM):
? 625
COMMAND? CS
STIFFNESS FACTORS CALCULATED
COMMAND? HE

COMMANDS AVAILABLE ARE:

IF:  INPUT FOOTING DIMENSIONS
IS:  INPUT SOIL PROPERTIES
IL:  INPUT LOADS ON THE FOOTING
ID:  INPUT DISPLACEMENTS OF FOOTING
CS:  CALCULATE STIFFNESS FACTORS
CD:  CALCULATE DISPLACEMENTS
CL:  CALCULATE LOADS
HE:  PRINT THIS HELP INFORMATION
EX:  EXIT

COMMAND? CD
VERTICAL DISPLACEMENT =        .177264  M
HORIZONTAL DISPLACEMENT =      .344398E-01  M
ROTATION =                     .192143E-01  RADIANS
COMMAND? IS
ENTER YOUNGS MODULUS FOR SOIL (KN/M2):
? 3450
ENTER POISSONS RATIO FOR SOIL:
```

```
? 0.2
COMMAND? CS
STIFFNESS FACTORS CALCULATED
COMMAND? CD
VERTICAL DISPLACEMENT =          .187003  M
HORIZONTAL DISPLACEMENT =        .336606E-01  M
ROTATION =                       .02027  RADIANS
COMMAND? EX
```

Program notes

(1) A glance at the program in Example 7.4 shows that it is written in quite a different style from those in the rest of this book. The program is used to illustrate an important programming technique of using a 'menu' of commands which direct the computer to carry out any of several simple tasks.
(2) The core of the program consists of lines 60 to 170 in which the prompt COMMAND? is given and a two letter reply expected. The program recognizes nine commands, one of which (EX) just stops the program, while the others each cause a given subroutine (at lines 1000, 2000 etc) to be executed. Any other command is ignored. The user can execute the subroutines in any order simply by giving the appropriate commands. Note the use of a very useful command HE (short for Help) which causes a list of the available commands to be printed. For convenience this routine is called automatically when the program is started.
(3) The program is based on Equations (7.15) to (7.17). These may be rewritten

$$V = \frac{2RE}{(1-\nu^2)}\delta_v = F_1\delta_v \tag{7.27}$$

$$H = \frac{4RE}{(1+\nu)(2-\nu)}\delta_h = F_2\delta_h \tag{7.28}$$

$$M = \frac{4R^3E\,\theta}{3(1-\nu^2)} = F_3\,\theta \tag{7.29}$$

where R is the radius of the footing and F_1, F_2, F_3 are stiffness factors.
(4) Commands IF and IS are used to input footing dimensions and soil properties. Command CS can then be used to calculate the stiffness factors.
(5) Command IL can be used to input loads V, H and M, and command CD then used to obtain corresponding displacements, or ID used to input displacements and CL to calculate the loads.
(6) The footing radius or soil properties could be changed, using IF or IS, and more calculations carried out, not forgetting to recalculate the stiffness factors by CS.

Example 7.5 Surface settlements due to vertical loading

Write a program to calculate the settlements at any point on the surface of an elastic soil mass due to a number of point loads and uniformly distributed vertical loads, at arbitrary x and y co-ordinates in plan.

The settlements due to a point load may be expressed explicitly, the surface settlement at a point a distance D from a vertical point load Q being given by

$$\delta_v = \frac{Q(1-\nu^2)}{\pi ED} \tag{7.30}$$

Note that the settlement is theoretically infinite immediately under the load, where $D = 0$.

For a uniform load over a circular area such an expression cannot be obtained. Instead a series of values of settlement factors F have been determined numerically for different values of r/r_c where r is the distance of the settlement point from the centre of circle of radius r_c carrying a uniform load/unit area p. The settlement is given by

$$\delta_v = \frac{p(1-\nu^2)r_c F}{E} \tag{7.31}$$

Suitable values of F may be stored in the program.

```
LIST

10      REM EXAMPLE 7.5. CALCULATION OF SURFACE SETTLEMENT PROFILES FROM
20      REM IMPOSED POINT AND CIRCULAR LOADINGS. USES APPROXIMATE FIT TO
30      REM DISPLACEMENT PROFILE FOR CIRCULAR LOADING
40      REM
50      REM DISPLACEMENT FACTORS FOR CIRCULAR LOADING
60      DATA 2.0, 1.97987, 1.91751, 1.80575, 1.62553, 1.27319
70      REM
80      DATA 1.27319, .51671, .33815, .252, .20045, .16626, .14315, .12576
90      REM
100     DIM F1(5),F2(8)
110     DIM XP(20),YP(20),PP(20),XC(20),YC(20),PC(20),XS(20),YS(20)
120     DEF FND(X1,Y1,X2,Y2)=SQR((X1-X2)^2 + (YI-Y2)^2)
130     REM
140     REM READ IN DISPLACEMENT FACTORS
150     REM
160     FOR I=0 TO 5
170     READ F1(I)
180     NEXT I
190     FOR I=1 TO 8
200     READ F2(I)
210     NEXT I
220     REM
230     REM ENTER DATA
240     REM
250     PRINT "ENTER SOIL PROPERTIES:"
260     PRINT "YOUNGS MODULUS (KN/M2):"
270     INPUT E
```

```
280     PRINT "POISSONS RATIO:"
290     INPUT NU
300     PRINT "ENTER INFORMATION ABOUT LOADING:"
310     PRINT "ENTER NUMBER OF POINT LOADS:"
320     INPUT NP
330     IF NP=0 GOTO 400
340     PRINT "FOR EACH POINT LOAD ENTER:"
350     PRINT "X-COORD (M), Y-COORD (M), LOAD (KN)"
360     FOR I=1 TO NP
370     PRINT "POINT LOAD",I
380     INPUT XP(I),YP(I),PP(I)
390     NEXT I
400     PRINT "ENTER NUMBER OF CIRCULAR LOADS:"
410     INPUT NC
420     IF NC=0 GOTO 490
430     PRINT "FOR EACH CIRCULAR LOAD ENTER:"
440     PRINT "CENTRE X,Y-COORDS (M), RADIUS (M), LOAD INTENSITY (KN/M2)"
450     FOR I=1 TO NC
460     PRINT "CIRCULAR LOAD",I
470     INPUT XC(I),YC(I),RC(I),PC(I)
480     NEXT I
490     PRINT "ENTER NUMBER OF POINTS FOR SETTLEMENT CALCULATION:"
500     INPUT NS
510     PRINT "ENTER X,Y COORDS OF EACH SETTLEMENT POINT (M):"
510     FOR I=1 TO NS
530     PRINT "SETTLEMENT POINT",I
540     INPUT XS(I),YS(I)
550     NEXT I
560     REM
570     REM CALCULATE SETTLEMENTS
580     REM
590     PRINT " "
600     PRINT "CALCULATED SETTLEMENTS"
610     PRINT " "
620     PRINT "X-COORD (M)","Y-COORD (M)","SETTLEMENT (M)"
630     PRINT " "
640     FOR I=1 TO NS
650     S = 0
660     IF NP=0 GOTO 710
670     FOR J=1 TO NP
680     D = FND(XP(J),YP(J),XS(I),YS(I))
690     S = S + PP(J) * (1 - NU*NU) / (3.14159 * E * D)
700     NEXT J
710     IF NC=0 GOTO 860
720     FOR J=1 TO NC
730     D = FND(XC(J),YC(J),XS(I),YS(I))
740     R = D / RC(J)
750     IF R>8 GOTO 830
760     IF R>1 GOTO 800
770     IR = INT(5 * R)
780     F = F1(IR) + (F1(IR+1) - F1(IR)) * 5 * (R - IR)
790     GOTO 840
800     IR = INT(R)
810     F = F2(IR) + (F2(IR+1) - F2(IR)) * (R - IR)
820     GOTO 840
830     F = 1 / R
840     S = S + PC(J) * (1 - NU*NU) * RC(J) * F / E
850     NEXT J
860     PRINT XS(I),YS(I),S
870     NEXT I
880     STOP
890     END

READY

RUN

ENTER SOIL PROPERTIES:
YOUNGS MODULUS (KN/M2):
? 3000
```

```
POISSONS RATIO:
? .2
ENTER INFORMATION ABOUT LOADING:
ENTER NUMBER OF POINT LOADS:
? 2
FOR EACH POINT LOAD ENTER:
X-COORD (M), Y-COORD (M), LOAD (KN)
POINT LOAD      1
? 0, 0, 200
POINT LOAD      2
? 10, 0, 300
ENTER NUMBER OF CIRCULAR LOADS:
? 2
FOR EACH CIRCULAR LOAD ENTER:
CENTRE X,Y-COORDS (M), RADIUS (M), LOAD INTENSITY (KN/M2)
CIRCULAR LOAD  1
? 0, 10, 2, 30
CIRCULAR LOAD  2
? 10, 10, 5, 30
ENTER NUMBER OF POINTS FOR SETTLEMENT CALCULATION:
? 5
ENTER X,Y COORDS OF EACH SETTLEMENT POINT (M):
SETTLEMENT POINT              1
? 1, 1
SETTLEMENT POINT              2
? 2, 2
SETTLEMENT POINT              3
? 5, 5
SETTLEMENT POINT              4
? 9, 9
SETTLEMENT POINT              5
? 25, 25

CALCULATED SETTLEMENTS

X-COORD (M)   Y-COORD (M)   SETTLEMENT (M)

 1              1             .123782
 2              2             .668757E-01
 5              5             .588818E-01
 9              9             .396679E-01
 25             25            .109683E-01
```

Program notes

(1) After inputting the data the program consists of several loops. The main loop is over the number of points at which settlements are required. Within this loop are two loops to add up the effects of settlement due to each of the point loads and each of the distributed circular loads.

(2) The program often has to calculate the distance between two points (x_1,y_1) and (x_2,y_2), so this operation is defined as a function at line 120.

(3) The program reads from data statements the values of F for values of r/r_c at intervals of 0.2 from 0 to 1.0 and at intervals of 1.0 from 1.0 to 8.0. These are stored in two arrays F_1 and F_2.

(4) When the program encounters an r/r_c value less than 1.0 it uses linear interpolation between the values at each interval of 0.2. Between 1.0 and 8.0 it uses a similar interpolation with the wider interval. For r/r_c values greater than 8.0 the approximation $F = r_c/r$

is used. Note that this corresponds exactly to considering the whole of the load to be concentrated at the central point. As long as one is far enough away the effect of a distributed load is the same as that of a point load.
(5) The method of carrying out the linear interpolation is of interest. The next point inside the r/r_c value is first identified by taking the integer part of r/r_c (or $5r/r_c$ if $r/r_c < 1.0$) and then the settlement factor calculated by adding to the settlement factor at the next point inwards a proportion of the difference between that value and the next value outwards. The proportion varies linearly from 0.0 at the inner point to 1.0 as the outer point is reached and is obtained by subtracting the integer part of r/r_c from r/r_c (or the integer part of $5r/r_c$ from $5r/r_c$ for the case $r/r_c < 1.0$).

Example 7.6 Cost analysis of a rectangular pile group

Write a program to assess the cost of a design for a rectangular piled foundation. The design of the foundation should include details of the dimensions of the foundation and the total applied load, and the length and diameter of the piles. The pile capacity is to be assessed using end bearing and side friction terms, including an α factor for the side friction. The soil is to be treated as an undrained clay with strength increasing with depth.

The cost of the foundation is to be made up from a lump sum for each pile required, plus a cost per metre length of pile.

```
LIST

10      REM EXAMPLE 7.6. ANALYSIS OF RECTANGULAR PILE GROUP TO ASSESS COSTS
20      REM
30      PRINT "COST ANALYSIS OF RECTANGULAR GRID OF PILES"
40      PRINT " "
50      PRINT "ENTER PROPERTIES FOR SOIL:"
60      PRINT " "
70      PRINT "UNDRAINED STRENGTH AT SOIL SURFACE (KN/M2)"
80      INPUT CO
90      PRINT "RATE OF INCREASE OF STRENGTH WITH DEPTH (KN/M3)"
100     INPUT C1
110     PRINT " "
120     PRINT "ENTER INFORMATION FOR FOOTING:"
130     PRINT " "
140     PRINT "BREADTH (M)"
150     INPUT BG
160     PRINT "LENGTH (M)"
170     INPUT LG
180     PRINT "TOTAL LOAD ON FOOTING (KN)"
190     INPUT PG
200     PRINT " "
210     PRINT "ENTER INFORMATION FOR PILES:"
220     PRINT " "
230     PRINT "PILE DIAMETER (M)"
240     INPUT D
250     PRINT "PILE ADHESION ALPHA FACTOR"
260     INPUT A
270     PRINT "LENGTH OF PILES (M)"
280     INPUT L
```

```
290     PRINT " "
300     PRINT "ENTER COSTING INFORMATION:"
310     PRINT " "
320     PRINT "MOBILISATION COST PER PILE"
330     INPUT CM
340     PRINT "COST PER METRE OF PILE"
350     INPUT CP
360     REM
370     REM CALCULATION OF NUMBER OF PILES AND COST, PRINT OUTPUT
380     REM
390     EB = 9 * (3.14159 * D * D * 0.25) * (C0 + C1 * L)
400     SF = A * D * 3.14159 * L * (C0 + C1 * (L * 0.5))
410     PC = EB + SF
420     N = INT(PG / PC) + 1
430     SM = SQR(LG * BG / N)
440     NL = INT(LG / SM) + 1
450     IF NL<2 THEN NL = 2
460     NB = INT(BG / SM) + 1
470     IF NB<2 THEN NB = 2
480     NP = NL * NB
490     SL = (LG - D) / (NL - 1)
500     SB = (BG - D) / (NB - 1)
510     C = CP * L
520     TP = CM + C
530     TT = TP * NP
540     F = 1 + 0.2 * L / BG
550     IF F>1.5 THEN F = 1.5
560     F = F * 5 * (1 + 1.2 * BG / LG)
570     GF = (BG*LG * F * (C0+C1*L)) + (2 * (BG+LG) * L * (C0+0.5*C1*L))
580     GP = PC * NP
590     F1 = GP / PG
600     F2 = GF / PG
610     PRINT " "
620     PRINT "ANALYSIS OF RECTANGULAR PILE GROUP"
630     PRINT " "
640     PRINT "SOIL PROPERTIES:"
650     PRINT "     UNDRAINED STRENGTH AT SURFACE (KN/M2)   ";C0
660     PRINT "     INCREASE OF STRENGTH WITH DEPTH (KN/M3)";C1
670     PRINT "PILE DATA:"
680     PRINT "     PILE LENGTH (M)                         ";L
690     PRINT "     DIAMETER (M)                            ";D
700     PRINT "     END BEARING (KN)                        ";EB
710     PRINT "     SIDE FRICTION (KN)                      ";SF
720     PRINT "     TOTAL CAPACITY (KN)                     ";PC
730     PRINT "PILE GROUP DATA:"
740     PRINT "     BREADTH X LENGTH (M)                    ";BG;"X";LG
750     PRINT "     TOTAL LOAD (KN)                         ";PG
760     PRINT "     MINIMUM NUMBER OF PILES REQUIRED        ";N
770     PRINT "     GRID BREADTH X LENGTH                   ";NB;"X";NL
780     PRINT "     TOTAL NUMBER OF PILES                   ";NP
790     PRINT "     SPACING OF PILES, BREADTH X LENGTH (M) ";SB;"X";SL
800     PRINT "SAFETY FACTORS:"
810     PRINT "     BASED ON TOTAL CAPACITY OF PILES        ";F1
820     PRINT "     BASED ON OVERALL STABILITY ANALYSIS     ";F2
830     PRINT "COST INFORMATION:"
840     PRINT "     COST OF EACH PILE                       ";TP
850     PRINT "     TOTAL COST OF PILE GROUP                ";TT
860     REM
870     REM OPTIONS ON CHANGING DATA
880     REM
890     PRINT " "
900     PRINT "CHANGE DATA (Y/N)";
910     INPUT A$
920     IF A$="N" GOTO 1250
930     IF A$<>"Y" GOTO 900
940     PRINT " "
950     PRINT "ENTER CODE:"
960     PRINT " "
970     PRINT "1 CHANGE C0              6 CHANGE D"
```

```
980     PRINT "2 CHANGE C1                 7 CHANGE A"
990     PRINT "3 CHANGE BG                 8 CHANGE L"
1000    PRINT "4 CHANGE LG                 9 CHANGE CM"
1010    PRINT "5 CHANGE PG                10 CHANGE CP"
1020    INPUT I
1030    PRINT "ENTER NEW VALUE:"
1040    ON I GOTO 1050,1070,1090,1110,1130,1150,1170,1190,1210,1230
1050    INPUT CO
1060    GOTO 390
1070    INPUT C1
1080    GOTO 390
1090    INPUT BG
1100    GOTO 390
1110    INPUT LG
1120    GOTO 390
1130    INPUT P
1140    GOTO 390
1150    INPUT D
1160    GOTO 390
1170    INPUT A
1180    GOTO 390
1190    INPUT L
1200    GOTO 390
1210    INPUT CM
1220    GOTO 390
1230    INPUT CP
1240    GOTO 390
1250    STOP
1260    END

READY

RUN

COST ANALYSIS OF RECTANGULAR GRID OF PILES

ENTER PROPERTIES FOR SOIL:

UNDRAINED STRENGTH AT SOIL SURFACE (KN/M2)
? 22
RATE OF INCREASE OF STRENGTH WITH DEPTH (KN/M3)
? 3.1

ENTER INFORMATION FOR FOOTING:

BREADTH (M)
? 4
LENGTH (M)
? 6
TOTAL LOAD ON FOOTING (KN)
? 3950

ENTER INFORMATION FOR PILES:

PILE DIAMETER (M)
? .4
PILE ADHESION ALPHA FACTOR
? .45
LENGTH OF PILES (M)
? 10

ENTER COSTING INFORMATION:

MOBILISATION COST PER PILE
? 470
COST PER METRE OF PILE
? 55

ANALYSIS OF RECTANGULAR PILE GROUP
```

```
SOIL PROPERTIES:
    UNDRAINED STRENGTH AT SURFACE (KN/M2)      22
    INCREASE OF STRENGTH WITH DEPTH (KN/M3)    3.1
PILE DATA:
    PILE LENGTH (M)                            10
    DIAMETER (M)                               .4
    END BEARING (KN)                           59.9415
    SIDE FRICTION (KN)                         212.057
    TOTAL CAPACITY (KN)                        271.999
PILE GROUP DATA:
    BREADTH X LENGTH (M)                       4 X 6
    TOTAL LOAD (KN)                            3950
    MINIMUM NUMBER OF PILES REQUIRED           15
    GRID BREADTH X LENGTH                      4 X 5
    TOTAL NUMBER OF PILES                      20
    SPACING OF PILES, BREADTH X LENGTH (M)     1.2 X 1.4
SAFETY FACTORS:
    BASED ON TOTAL CAPACITY OF PILES           1.37721
    BASED ON OVERALL STABILITY ANALYSIS        6.24608
COST INFORMATION:
    COST OF EACH PILE                          1020
    TOTAL COST OF PILE GROUP                   20400

CHANGE DATA (Y/N)? Y

ENTER CODE:

1 CHANGE CO             6 CHANGE D
2 CHANGE C1             7 CHANGE A
3 CHANGE BG             8 CHANGE L
4 CHANGE LG             9 CHANGE CM
5 CHANGE PG            10 CHANGE CP
? 8
ENTER NEW VALUE:
? 14

ANALYSIS OF RECTANGULAR PILE GROUP

SOIL PROPERTIES:
    UNDRAINED STRENGTH AT SURFACE (KN/M2)      22
    INCREASE OF STRENGTH WITH DEPTH (KN/M3)    3.1
PILE DATA:
    PILE LENGTH (M)                            14
    DIAMETER (M)                               .4
    END BEARING (KN)                           73.9656
    SIDE FRICTION (KN)                         345.964
    TOTAL CAPACITY (KN)                        419.93
PILE GROUP DATA:
    BREADTH X LENGTH (M)                       4 X 6
    TOTAL LOAD (KN)                            3950
    MINIMUM NUMBER OF PILES REQUIRED           10
    GRID BREADTH X LENGTH                      3 X 4
    TOTAL NUMBER OF PILES                      12
    SPACING OF PILES, BREADTH X LENGTH (M)     1.8 X 1.86667
SAFETY FACTORS:
    BASED ON TOTAL CAPACITY OF PILES           1.27574
    BASED ON OVERALL STABILITY ANALYSIS        8.46218
COST INFORMATION:
    COST OF EACH PILE                          1240
    TOTAL COST OF PILE GROUP                   14880

CHANGE DATA (Y/N)? N
```

Program notes

(1) Lines 10 to 350 carry out the usual task of prompting the user for the input quantities.

(2) Lines 390 to 410 calculate the bearing capacity of an individual pile from the sum of end bearing (using an N_c value of 9) and side friction. Line 420 then calculates the minimum necessary number of piles.
(3) Since the piles are to be arranged on a rectangular grid, line 430 calculates the maximum spacing which can be used — by assuming that each pile carries a fraction $1/N$ of the total load and therefore supports an area which is a fraction $1/N$ of the total footing area. Lines 440 to 470 then calculate the number of piles along and across the foundation — subject to the arbitrary limit that at least two piles in each direction are to be used.
(4) The total number of piles is then the product of NL and NB, which of course gives slightly more than the minimum necessary number of piles N. At lines 490 and 500 the pile spacing is calculated in each direction (remembering to allow for a total of one diameter at the edge of the footing). The spacing is the centre-to-centre distance.
(5) The cost of each pile is totalled and the overall cost calculated at lines 510 to 530.
(6) It is normal design practice to take the capacity of a piled foundation as the smaller of:

(a) The sum total capacity of the individual piles
(b) The capacity of the whole footing treated as a deep block footing.

At lines 540 to 570 the overall capacity of the footing is determined by adding the capacity on the base of the footing as calculated from Equation (7.6) and the side friction on the footing, assuming full roughness.
(7) At lines 590 to 600 factors of safety are calculated for the cases of both the total pile capacity and the footing capacity.
(8) Lines 610 to 850 output an extensive amount of information about the calculation, as well as the total cost.
(9) Lines 860 to 1260 are used to provide a convenient option for changing data.
(10) The application used to illustrate the program shows a typical calculation. At first, piles 10 m long are chosen, giving a total cost of £20,400. When the pile length is increased to 15 m the number of piles is reduced sufficiently (from 20 to 12) for the cost to fall to £14,880.

PROBLEMS

(7.1) Modify the program in Example 7.1 to include footings carrying eccentric or inclined loads, using Equations (7.2) to (7.4) or one of the more complex bearing capacity formulae given in textbooks.
(7.2) Write a program to design square pad footings for individual column loads and known soil conditions. Use the form of the bear-

ing capacity formula appropriate to a square footing. Since the bearing pressure is dependent on the width of the footing (in the third term in Equation (7.1)) it will be necessary to use an iterative or trial-and-error procedure to determine the required foundation area and hence width. In practice the dimensions of footings on a particular building would be restricted to a few standardized sizes; one approach would therefore be to calculate the allowable loads for a set of such footings (incorporating a safety factor on the ultimate bearing capacity) and selecting a suitable footing in each case by comparing the actual column load with the set of allowable loads.

(7.3) Write a program, or modify the program in Example 7.3, to determine the stresses at any point under a strip footing carrying a linearly increasing loading using Equations (7.10) to (7.12).

(7.4) Combine the programs from Example 7.3 and Problem (7.3) and use the principle of superposition to enable the stresses under an embankment to be calculated.

(7.5) The program in Example 7.4 is not 'foolproof' in that if commands are not given in a sensible order then the program can deliver nonsensical results. For example, if after several calculations some new soil properties are given using the IS command and the user forgets to recalculate the stiffness factors using the CS command, then the old properties will still be used in any subsequent calculations. Make the program more 'robust' by removing the CS command and calculating stiffness factors automatically either

(a) the first time both IF and IS have been used to give sufficient information to calculate the factors

(b) any time IF and IS are used to change parameter values.

The most convenient way to do this may be to set some 'flags' which would initially have value zero, but are changed to 1 when IF or IS is used.

(7.6) Several other programs in this book could be rewritten using the 'menu' method of Example 7.4. Use this method to rewrite for instance one of the Coulomb analysis programs (5.2 or 5.3), using commands to input soil properties and wall and backfill geometry, and to carry out the main calculation. Note that although your new program may be much longer, it is often easier to use and easier to extend to more complex calculations.

(7.7) Develop the program from Problem (7.4) to calculate settlements of an embankment due to consolidation of an underlying clay stratum. Use the method outlined in section 7.5. The clay stratum should first be divided into layers, the thickness of each being relatively small compared with the width of the embankment. The program should then calculate and print out the initial and final stresses at the centre of each layer at a number of sections, say under the middle of the embankment, under its edge, and halfway between these two. Suitable values of m_v, determined from oedometer tests over appropriate stress ranges, may then be input and the compression of each layer and hence the total settlement calculated.

(7.8) Write a program to calculate the vertical stresses and settlement at the surface under the corner of a uniformly loaded rectangular area (see Figure 7.5 and Equations (7.20) and (7.21)). Develop this program using the principle of superposition as shown in Figure 7.6 to allow the stresses and settlement to be determined at any point under a uniformly loaded area made up of rectangular shapes. The program of Example 7.5 should prove a useful model.

(7.9) The program in Example 7.6 can be improved in a variety of ways; the following are some suggestions.

(a) Include a check that neither of the pile spacings SL or SB are less than the pile diameter D, and print a suitable warning message if this should occur.

(b) Print a warning message if the overall factor of safety for the footing is less than unity.

(c) The program sometimes chooses a grid of piles which is too pessimistic because of the way in which NB and NL are calculated. For instance, in the first part of the example given the minimum number of piles required is 15, but the program chooses a 4×5 grid. Clearly a 4×4 grid would be better, and a 3×5 grid the optimum. Either write some extra code to allow the programmer to override the grid chosen, or write a routine which checks whether a better grid would be possible (for instance by checking if (NB − 1)* NL is greater than N).

Index